John William Dawson

Handbook of geology for the use of Canadian students

John William Dawson

Handbook of geology for the use of Canadian students

ISBN/EAN: 9783337204433

Printed in Europe, USA, Canada, Australia, Japan

Cover: Foto ©berggeist007 / pixelio.de

More available books at **www.hansebooks.com**

HANDBOOK

OF

GEOLOGY

FOR THE USE OF

CANADIAN STUDENTS

By SIR J. WILLIAM DAWSON,

C.M.G., LL.D., F.R.S., ETC.

PRINCIPAL OF McGILL UNIVERSITY.

MONTREAL:

DAWSON BROTHERS, PUBLISHERS.

1889.

"WITNESS" PRINTING HOUSE,
MONTREAL.

PREFACE.

This work is intended to serve as lecture notes for teachers of Geology, more especially in the Dominion of Canada, and as a guide to Canadian Geology for private students, enquirers and travellers. The first part relates to the general principles of the science, with examples, as far as possible, from Canadian minerals and rocks. The second part gives an outline of Geological Chronology, illustrated by Canadian rock-formations and fossils. The third part gives details as to the Physical Geography and Geology of Canada. In this part reference is made to authorities and to works giving more detailed information; and credit is given, as far as possible, to the original observers of the more important geological facts.

SEPT., 1889.

TABLE OF CONTENTS.

PART II.—Historical Geology.

PART III.—Canadian Topography and Geology.

WESTERN CANADA.

GEOLOGICAL SHADING.

LAURENTIAN...........
L. SILURIAN...........
PALÆOZOIC (UNDETERMINED).

DEVONIAN............
CRETACEOUS..........
TERTIARY............

INTRODUCTORY.

GEOLOGY, or, as it has been sometimes termed, Geognosy, is the scientific knowledge of the earth; or more particularly of that rocky crust of the earth on which its superficial features depend, which affords to us mineral products and soils, on which animals and plants exist, and in which are preserved the monumental records of the changes which our planet has experienced in past time.

Previous to experience and observation, it might be supposed that all the rocks of the earth's crust are of the same age; but on careful study we find that this is not the case. On the contrary, the crust is built up of successive strata or beds deposited one over the other in such a manner that the upper are the newer and the lower the older. Thus we obtain a chronology of the rocks. Further, as the successive beds contain remains of animals and plants which lived at the times when these beds were deposited, we also obtain a chronology of animal and vegetable life. In connection with this succession, we can consider the causes of geological change, whether acting at present or in past time.

Geology may be studied with reference to its practical pursuit as a method of scientific investigation, or with reference to the theories of the earth deducible from its facts, or with reference to its applications to the arts of life. These several aspects of the subject may be termed—

    1. Practical Geology.

    2. Theoretical and Dynamical Geology.

    3. Applied Geology.

The first is that which should engage the attention of the student at the outset, as being preliminary to the successful cultivation of the others; but in studying it, reference may be made to its bearings on the second and third.

A

Practical geology may be arranged under the following heads:—

(1.) GENERAL.

I. LITHOLOGY or PETROGRAPHY—or the study of *Rocks* as mineral aggregates and as materials composing the earth's crust. This study is best carried on with the aid of properly named hand specimens of minerals and rocks, and is much aided by chemical tests and by the examination of sections of rocks under the microscope.

II. PALÆONTOLOGY—the study of the fossil remains of animals and plants imbedded in the earth's crust, in connection with the succession of deposits ascertained by stratigraphical investigation. This subject requires some preliminary knowledge of zoological and botanical classification, and is studied by comparison of museum specimens and by collecting and determining fossils.

III. STRATIGRAPHY or PETROLOGY—or the consideration of the arrangement of the rocky masses of the earth on the large scale. This study requires the aid of maps and sections of the structure of portions of the earth, and is carried on in nature by the examination of natural sections, in cliffs, quarries, mines, and other exposures of rocks.

(2.) HISTORICAL.

IV. HISTORICAL GEOLOGY is the application of the above to the geological history of the earth, and connects the elements of practical geology with the theory and applications of the subject.

The first, second and third of these subjects may be regarded as general and preliminary to the fourth, which includes geology proper. We shall therefore first consider the elements of Lithology, Palæontology and Stratigraphy, in so far as necessary, and then proceed to the main subject of Geological History, more especially in relation to Canada.

[In the regular University curriculum the student is supposed to have given some attention to the elements of Chemistry, Botany and Zoology. He is thus prepared in the ordinary course in Geology to enter on the study of Lithology, Palæontology, and Stratigraphy and in the honour course to go more fully into the determination of rocks and fossils, and into local stratigraphy and descriptive and theoretical geology.]

PART I. GENERAL GEOLOGY.

I. LITHOLOGY.*

(1.) CHEMISTRY OF ROCKS.

Of about seventy elements or simple substances known to chemistry, only sixteen enter into the composition of the more common rocks which constitute nearly the whole of the earth's crust. These are, in the order of their relative importance :—

Non-Metallic Elements.	Metallic Elements.
Oxygen.	Iron.
Silicon.	Aluminium.
Sulphur.	Calcium.
Chlorine.	Magnesium.
Carbon.	Sodium.
Hydrogen.	Potassium.
Fluorine.	Barium.
Phosphorus.	Manganese.

Of the above, only Oxygen, Sulphur, Carbon and Iron can exist in nature in a pure or uncombined state. The more common minerals are all compounds of two or more elements.

Oxygen is the most important element in the crust of the earth, since in the ordinary rocks the other elements almost always occur in combination with this, as oxides. Thus Silica or Flint is Oxide of Silicon, Alumina the earth of clay is Oxide of Aluminium, Lime is Oxide of Calcium. The ordinary ores of Iron are oxides of the metal.

Next to Oxygen the most important element is *Silicon*. Combining with Oxygen this forms Silica, and Silica has the property of combining with many other elements to form *Silicates*, which are the most common constituents of minerals and rocks. Of these Silicates the most abundant

---

* The term Petrography is sometimes employed for Lithology, but as the similar term Petrology is also applied by some geologists to Stratigraphy, and as both are applicable to rocks rather than to hand specimens, it is not so definite as Lithology.

are those of Aluminium, Calcium, Magnesium, Sodium and Potassium ; and these are variously combined and mixed with one another to constitute the more complex minerals and rocks.   Silicates sometimes contain water as an essential constituent, when they are termed *Hydrous Silicates.*

Other important Oxygen compounds are the *Carbonates, Sulphates* and *Phosphates.*   Thus Calcium Carbonate is common Limestone, Calcium Sulphate is Gypsum, and Calcium Phosphate is Apatite or Bone-earth.

Some important constituents of rocks are not Oxides, as Sodium Chloride or common Salt, Calcium Fluoride or Fluorspar, Iron Bi-Sulphide or Iron Pyrite.

There is a peculiar group of minerals and rocks of organic origin into which Carbon enters as a principal ingredient.   These are the Coals, Asphalt and Bitumen.

(2.) MINERALOGY OF ROCKS.

Of the chemical compounds above referred to, those which constitute the majority of rocks are the following :—

1. Quartz or Silica.	
2. Felspar.	
3. Mica.	
4. Hornblende.	
5. Pyroxene.	Anhydrous Silicates.
6. Garnet.	
7. Chrysolite.	
8. Nepheline.	
9. Leucite.	
10. Talc.	
11. Serpentine.	Hydrous Silicates.
12. Chlorite.	
13. Calcite.	
14. Dolomite.	
15. Gypsum.	Carbonates, Sulphate,
16. Apatite.	Phosphate, Fluoride,
17. Fluor Spar.	and Chloride.
18. Rock Salt.	
19. Magnetite.	
20. Hematite.	Oxides and Sulphide
21. Limonite.	of Iron.
22. Pyrite.	

23. Coal.

24. Bitumen and Asphalt. } Carbonaceous Minerals.

25. Graphite.

[For the characters and description of the more important of the minerals, see the appendix. If the learner has not previously studied Mineralogy, he should refer constantly to the descriptions of minerals unknown to him, and should have an opportunity to examine specimens.]

(3.) CLASSIFICATION AND STUDY OF ROCKS.

Some rocks, as quartz rock and limestone, are definite chemical compounds, and consist of one mineral species only; but even these are often mixed with foreign matters; and the greater part of rocks are mixtures of different mineral substances in various proportions. As these mixtures are regulated by no definite law of proportion, it follows that such rocks pass into each other by indefinite gradations. Hence the nomenclature and classification of rocks are attended with many difficulties.

For purposes of practical geology it is important to consider the classification of rocks under three aspects.

1. With reference to their *Origin*, rocks may be:—

(a) *Aqueous* or *Sedimentary*, that is, they may have been deposited as sediments, as sand, clay, &c., in water, and such deposition may have been aided or modified by accumulations of organic matter, as shells, corals, drifted plants, &c. Rocks of this kind are *Conglomerates*, or hardened gravels; *Sandstones*, or consolidated sand; *Shales*, which are compressed clay; *Limestones*, which are usually aggregates of shells, corals, &c., or fragments of these.

(b) *Igneous* or *Aqueo-igneous*—products of the action of heat in the interior of the earth. Of this kind are lavas, scoriæ, pumice, and volcanic ashes.

(c) *Metamorphic*—that is they may be sediments or volcanic beds which have been so modified by heat or pressure as to assume a crystalline condition accompanied, in many cases, by some chemical change. Examples of these are—Quartzite, produced by the hardening of Sandstone; Crystalline Schists, produced from altered clays: Marble, produced by the alteration of limestone; and Graphite and Anthracite, by that of coal.

2. With reference to their *Predominant Chemical Ingredients*, rocks may be regarded as (a) *Silicious*, (b) *Argillaceous*, (c) *Calcareous*, (d) *Carbonaceous*, (e) *Ferruginous*. The Silicious rocks, which are by far the most abundant, may further be divided into those that are *Acidic* or have

an excess of Silica, and those that are *Basic* or have an excess of the elements with which Silica is combined.

3. With reference to their *Texture*, rocks may be :—

(a) *Fragmental*, or *Clastic*, that is, composed of broken-up remains of older rocks. Of this kind are conglomerates, sandstone and clay.

(b) *Crystalline*, or composed of crystals of one or more minerals united together. Of this kind are granite and crystalline marble.

(c) *Organic*, or retaining the structure of organic bodies, as coral and crinoidal limestones, and coals.

The above grounds of classification are of course allied with each other. Thus fragmental rocks are for the most part aqueous. The crystalline rocks are for the most part of igneous or metamorphic origin, though some, like gypsum and rock salt, are aqueous. We may thus adopt one of the above arrangements as the dominant or general one, and use the others in subordination to it; and the first consideration or that of *origin* is probably at present the most available for the larger groups. Our general division of rocks may therefore be as follows :—

Class I.
IGNEOUS ROCKS. } including { (1) *Volcanic* or *Superficial*.
(2) *Plutonic* or *Nether*.

Class II.
AQUEOUS ROCKS. } including { (1) *Unaltered*.
(2) *Altered* or *Metamorphic*.

The Plutonic rocks correspond in the main with the "Intrusive" rocks, properly so called, and the Volcanic with the "Effusive" or lavas proper. The Plutonic rocks are in structure holocrystalline. The Volcanic are usually partly vitreous or compact.

CLASS I.—IGNEOUS ROCKS.

Section 1. VOLCANIC.

These are superficial products of Igneous action. All of them are Silicates, having usually Aluminium, Calcium and Magnesium as the principal bases. They may be divided into sub-sections, in accordance with the proportions of acid and base; those containing less than 60 per cent. of Silica, being called Basic, and those with more than 60 per cent. Acidic. Some lithologists also distinguish Intermediate and Sub-acid rocks having 60 to 65 per cent. of Silica, and Ultra-basic rocks containing less than 45 per cent.

Sub-Section 1. BASIC VOLCANIC ROCKS.

Doleritic Lava is poured forth in a molten state by modern volcanoes, and consists of Pyroxene with basic Felspars. It generally presents a vesicular or scoriaceous appearance, caused by the expansion of included vapours and gases, and it has usually a dark colour, becoming reddish or yellow on weathering, owing to the presence of iron in low states of oxidation, (Protoxide, Magnetite and Ilmenite.) In ancient lavas the vesicles often become filled by aqueous infiltration with various minerals, when the texture of the rock is said to be *Amygdaloidal*.

Basalt is a dark-coloured finely crystalline compact lava, which often exhibits columnar structure. *Melaphyre* is a variety more crystalline than Basalt, and presents a transition to the true Diabase of the Plutonic group. *Andesite* is a variety containing Hornblende. Leucitic and Peridotic lavas contain crystals of Leucite and Olivine.

Sub-section 2. ACIDIC VOLCANIC ROCKS.

Trachytic Lava is a light-coloured lava containing an excess of Silica, and produced by volcanoes in the same manner with ordinary lava. It is vesicular, and when highly so passes into *Pumice*. The name *Rhyolite* is given to those highly acidic Trachytes having an excess of quartz.

Trachyte is a more compact rock of the same character, consisting chiefly of orthoclase, usually with a little hornblende and mica. When quartz is present it becomes *Quartz-trachyte*. It is more or less finely crystalline, and sometimes has imbedded crystals of orthoclase felspar, giving it the texture known as *porphyritic*.

Obsidian and *Pitchstone* are volcanic glasses of similar composition to trachyte but vitreous in texture.

Volcanic Agglomerate and *Volcanic Tuff* are fragmental deposits made up of the stones and dust ejected from volcanic orifices. Their materials may either be those of the basic or acidic lavas or a mixture of both. They are strictly volcanic rocks, though their materials are often arranged in beds and subsequently consolidated by the action of water.

Section 2. PLUTONIC.

These are the nether or underlying products of igneous action. Being slowly cooled they are more highly crystalline than the rocks of the previous section, and having been consolidated at great depths below the surface, they do not become visible till after the removal of the more superficial volcanic products. Hence the rocks of this section visible at the surface, are usually of greater age than the volcanic rocks.

Sub-section 1. BASIC PLUTONIC ROCKS.

Diabase consists of the same material with Doleritic lava and Basalt, and passes into it; but is usually much more coarsely crystalline. It often contains hydrous minerals, as chlorite. *Gabbro* is a name applied to a variety of Diabasic rock in which Pyroxene occurs in the form of Diallage. The *Peridotites* are ultra-basic Plutonic rocks with a large proportion of Olivine.

Diorite or Greenstone is a crystalline mixture of Hornblende, usually dark coloured or greenish, with a triclinic felspar. This and the previous rock present great varieties of coarse and fine crystallization. The Diorites with fibrous hornblende are supposed to be altered Dolerites. Those with massive and usually dark-coloured hornblende are the true Diorites.

Syenite is a crystalline mixture of Hornblende and Orthoclase or Potash Felspar. It may be regarded as a sub-acid rock, and by the addition of Quartz it becomes an caidic rock and passes into Hornblendic Granite.

Sub-section 2. ACIDIC PLUTONIC ROCKS.

Granite is a crystalline mixture of Felspar (Orthoclase, or Albite) with Quartz and Mica. It may be coarse or fine grained, and sometimes becomes porphyritic by the admixture of large felspar crystals. *Hornblendic* or *Syenitic Granite* contains Hornblende with or instead of Mica. *Protogine* contains Talc as well as Mica. *Graphic Granite* is a binary variety found in veins. It is destitute of mica, and has the quartz arranged in plates in accordance with the cleavage of the felspar.

Felsite is a hard, finely crystalline or compact mixture of Felspar and Quartz. It is sometimes called *Petrosilex* and *Felstone*. When distinct crystals of orthoclase Felspar are developed in it the porphyritic texture is produced. This is ordinary or *Felsite porphyry*, but other igneous rocks may assume the porphyritic structure.

The above are only a few of the more ordinary igneous rocks, which should be known to the student by specimens, and if possible also by their microscopic structure.

The nature of igneous rocks leads us to inquire as to the deep-seated sources, under the superficial crust, from which they come. The fact that igneous rocks are very generally distributed over the earth's surface, so that few large regions are destitute of them, proves that the sources of such material must be very widely spread, if not universal. The further fact that igneous rocks are in all parts of the world of the same general character, shows in like manner that their origin is in general and uniform "magmas" or pasty sheets of more or less uniform heated and plastic matter, spread under the crust. Lastly, the fact that in every region the igneous rocks are of both groups, the basic and the acidic, shows that both kinds of material exist, and to some extent separate from each other, under all parts of the crust. These great facts have important bearings on our notions of the interior of the earth, and on the origin of igneous rocks, which may be formulated as follows :—

(1.) Since the dawn of geological science, it has been evident that the crust on which we live must be supported on a plastic or partially liquid mass of heated rock, approximately uniform in quality under the whole of its area. This is a legitimate conclusion from the wide distribution of volcanic phenomena, and from the fact that the ejections of volcanoes, while locally of various kinds, are similar in every part of the world. It led to the old idea of a fluid interior of the earth, but this is now generally abandoned, and this interior heated and plastic layer is regarded as merely an under-crust.

(2.) We have reason to believe, as the result of astronomical investigations, that, notwithstanding the plasticity or liquidity of the under-crust, the mass of the earth—its nucleus as we may call it—is practically solid and of great density and hardness. Thus we have the apparent paradox of a solid yet fluid earth; solid in its astronomical relations, liquid or plastic for the purposes of volcanic action and superficial movements.

Bonney has also suggested the important consideration that a mass may be slowly mobile under long continued pressure, while yet rigid with reference to more sudden movements. An objection has been taken to the effect that the supposed ellipsoidal form of the equator is inconsistent with a plastic sub-crust. But the existence of this ellipsoidal form is not absolutely certain, or, if it exists, the divergence from the circular form is very minute.*

(3.) The plastic sub-crust is not in a state of dry igneous fusion, but in that condition of aqueo-igneous or hydro-thermic fusion which arises from the action of heat on moist substances, and which may either be regarded as a fusion or as a species of solution at a very high temperature. This we learn from the phenomena of volcanic action, and from the composition of the volcanic and plutonic rocks, as well as from such chemical experiments as those of Daubrée, and of Tilden and Shenstone.

(4.) The interior sub-crust is not perfectly homogeneous, but may be roughly divided into two layers or magmas ; an upper, highly siliceous or acidic, of low specific gravity, and light-coloured, and corresponding to such kinds of plutonic and volcanic rocks as granite and trachyte ; and a lower, less siliceous or more basic, more dense, and more highly charged with iron, and corresponding to such igneous rocks as the dolerites, basalts, and kindred lavas. It is interesting here to note that this conclusion, elaborated by Durocher and Von Walterhausen, and usually connected with their names, appears to have been first announced by John Phillips, in his " Manual of Geology," and as a mere common-sense deduction from the observed phenomena of volcanic action, and the probable results of the gradual cooling of the earth. It receives striking confirmation from the observed succession of acidic and basic volcanic rocks of all geological periods, and in all localities. It would even seem, from recent spectroscopic investigations of Lockyer, that there is evidence of a similar succession of magmas in the heavenly bodies, and the discovery by Nordenskiold of native iron in Greenland basalts, affords a probability that the inner magma is in part metallic. †

---

* Hopkins, Mallet, Sir William Thomson, and Prof. G. H. Darwin maintain the solidity and rigidity of the earth on astronomical grounds ; but different conclusions have been reached by Hennesey, Delaunay, and Airy. In America, Hunt, Barnard and Crosby, Dutton, LeConte and Wadsworth have discussed these questions.

† These basalts occur at Ovifak, Greenland. Andrews has found small particles of iron in British basalts. Prestwich and Judd have referred to the bearing on general geology of these facts, and of Lockyer's suggestions.

(5.) Where rents or fissures form in the upper crust, the material of the lower crust is forced upwards by the pressure of the less supported portions of the former, giving rise to volcanic phenomena either of an explosive or quiet character, as may be determined by contact with water. The underlying material may also be carried to the surface by the agency of heated water, producing those quiet discharges which Hunt has named crenitic. There can be little doubt that the weight of the crust pressing downward on the interior magmas, especially in places where cracks or folds have caused unequal pressure, is the cause of the ejection of molten rocks, which may thus be forced upward into fractures, forming dykes, or pressed between the beds, forming igneous floors or *Laccoliths*, or may be raised to the craters of volcanoes and caused to flow over as currents of lava. All this may go on quietly, but where water in a superheated state is intimately mixed with the molten rock, and is prevented by pressure from passing into vapour, or when water is introduced into the ascending lava from porous beds, or from fissures communicating with the sea, the sudden evaporation of the water produces tremendous explosions. Thus gravitative pressure is the cause of quiet ejections of volcanic products, the explosion of steam is the cause of the more violent phenomena. It is to be observed here that explosive volcanic phenomena, and the formation of cones, are, as Prestwich has well remarked, characteristic of an old and thickened crust; quiet ejection from fissures and hydro-thermic action may have been more common in earlier periods and with a thinner over-crust.

Active volcanic phenomena are not now manifested within the Dominiom of Canada ; but igneous rocks of all ages from the Laurentian to the Pliocene exist in various portions of its area, and will be noticed in connection with the aqueous rocks of the periods to which they belong, and with reference to the classification and principles above stated.

Igneous rocks are to be studied with reference to their crystalline, vitreous or fragmental condition, their various constituent and accidental minerals, the changes to which they have been subjected by subsequent decomposition, or from the injection of aqueous materials in solution, and the crushing or lamination which may have been induced by pressure after their consolidation.

CLASS II.—AQUEOUS ROCKS.

Section 1. UNALTERED AQUEOUS ROCKS.

These may be produced either by the mechanical distribution of sediment in water, by chemical precipitation, or by the accumulation of the remains of animals and plants.   Processes of these kinds are now going on, and must have been in operation throughout geological time.   Crystalline rocks have been undergoing decay, whereby their grains and crystals of quartz have been separated as sand, while their felspathic or hornblendic material has been decomposed into clay, marl, &c.   The surf on coasts, and running water on the land, have been grinding rocks into pebbles ; corals and shells have been accumulating as beds of limestone in the sea.   By such processes the immense sheets of aqueous rock spread over our continents have been slowly accumulated, so that they now present an aggregate thickness which has been estimated at 70,000 feet, or more.   The ways in which they have been elevated from their original sub-aqueous position, disturbed from their horizontal attitude, and hardened and altered, will be considered under a subsequent heading. The principal kinds are the following :—

Conglomerate consists of pebbles of hard, usually silicious, rocks, united by a paste or cement which may be silicious, argillaceous, calcareous or ferruginous. Conglomerates are beds of gravel, and they indicate the somewhat powerful action of water as an abrading and removing agent.   They have often been formed along old lines of coast, and are consequently irregular in their bedding and limited in their horizontal distribution.   The terms *Volcanic Breccia* and *Agglomerate* are applied to rocks composed of angular fragments.   Volcanic agglomerate has already been referred to ; but besides this, Breccias are accumulated by aqueous agencies in caves and fissures, and are also derived from the debris of hard rocks disintegrated by frost, and spread out by water without rounding the edges of the fragments.

Grit is a rock composed of coarse sand or small stones, and is intermediate between the last rock and the next.

Sandstone is composed of grains of sand, more or less fine, and either angular or rounded, cemented together.   When mixed with clay it becomes argillaceous sandstone.   When cemented by carbonate of lime it becomes calcareous sandstone.   Its grains are often superficially stained of red or brown colours by the oxide of iron. *Freestone* is a term applied to the softer and more easily worked sandstones ; *Flagstone* to the laminated varieties.   The harder varieties pass into *Quartzite.* *Greensand* is a variety coloured by grains of the hydrous silicate named glauconite.   Sandstones with the surfaces of bedding and lamination covered with plates of mica are *micaceous* sandstones.

Shale is hardened clay or mud, having a laminated texture, due either to original deposition in layers or to subsequent pressure.   On the one hand it passes into soft clay, on the other by metamorphism into slate.   *Arenaceous shale* is mixed with fine sand and passes into sandstone.   *Carbonaceous shale* is mixed and blackened with coaly matter.   *Bituminous shale* or *Pyroschist* is impregnated with bituminous matter.

Calcareous shale contains limestone in a fine state of division, and effervesces with an acid. *Fireclay* is a soft variety rendered infusible by the absence of alkaline matter. It is often associated with beds of coal. *Kaolin* is a fine clay resulting from the decomposition of felspar. *Loess* is the alluvial mud deposited in lakes and rivers. *Loam* is a mixture of sand and clay.

Conglomerates, sandstones and shales are typically clastic or fragmental rocks, and in studying them we have to consider the nature and origin of their constituent pebbles and grains, the amount of abrasion or decomposition to which these have been subjected, the nature of the cementing material, if any, which has been deposited in their interstices and binds them together, and the changes which they have experienced from heat or pressure.

Limestone includes all the unaltered rocks composed of calcium carbonate, or calcite. It is distinguished by its softness as compared with quartz and most of the silicious stones, and by effervescing with an acid. It may be earthy, compact, crystalline, massive or laminated in structure ; or with reference to matters mixed with it, argillaceous, bituminous, ferruginous, or cherty. *Oolite* is a variety composed of minute rounded concretions, which often show under the microscope a radiating prismatic structure as well as concentric lamination. *Travertin* or *Calcareous Tufa* is a limestone deposited by calcareous springs. *Stalactite* and *Stalagmite* are similar matter deposited on the roofs and floors of caverns. By mixture with fragments of limestone or of bone, Stalagmite may become a *calcareous* or *bone Breccia.*

Coral and *Shell Limestone* and *Crinoidal Limestone,* or more generally *Organic Limestones,* are composed of fragments of calcareous organisms, sometimes apparent to the eye, in other cases visible only under the microscope. *Chalk* is an organic limestone made up of tests of Foraminifera mixed with the minute organic bodies named Coccoliths.

Dolomite is a double calcium and magnesium carbonate. It may be distinguished from common limestones by its higher lustre, slightly greater weight, failure to effervesce with cold acid, and by often weathering of a rusty colour, in consequence of the presence in it of ferrous carbonate.

Marl is an earthy mixture of calcium carbonate with clay or sand. The calcareous matter is sometimes in a fine state of division and sometimes as fragments of shells (shell marl). Marl is distinguished from ordinary clay by effervescing briskly when treated with an acid.

Gypsum, or Calcium Sulphate, is of less common occurrence than limestone, but sometimes constitutes thick beds of great purity. *Anhydrite* is often associated with the ordinary hydrous variety.

Coal and carbonaceous rocks will be referred to under the heading of Minerals in the Appendix.

Iron ores will also be noticed under the same heading.

Section 2.   METAMORPHIC ROCKS.

These are rocks originally aqueous or aqueo-igneous, which have been subjected to the action of heat and pressure, along with chemical agencies, until their particles have so rearranged themselves as to give a crystalline character accompanied by differences in the state of combination of the contained elements.

The metamorphic rocks are intermediate in character between the unaltered aqueous and the plutonic series. On the one hand they pass into ordinary aqueous rocks, on the other by becoming highly crystalline and losing their original bedding, they graduate into plutonic rocks. The principal varieties of these metamorphosed rocks are the following :—

Quartzite or *Quartz Rock* is a result of the alteration of sandstone, whereby its grains of sand become inseparable and sometimes indistinguishable.

Gneiss is a product of the alteration of sediments containing sufficient basic matter for the production of Felspar and Hornblende or mica. It thus resembles granite in composition, and is distinguished by its laminated structure and stratified arrangement. Many gneisses may have originally been bedded trachytes or volcanic tuffs.

Argillite or *Clay Slate* is a product of the alteration and hardening of clay or shale. It is remarkable for the development in it of *slaty structure*, which arises from the forcing by lateral pressure, of all flat particles in a soft mass into positions in which they lie at right angles to the direction of pressure. In this way the most perfect lamination is often produced in planes quite different from those of bedding.

Mica Schist is a crystalline mixture of Quartz and Mica. It is a product of the alteration of shales. It often contains disseminated minerals, as pyrite, garnet, staurolite, or chiastolite. By addition of felspar it passes into gneiss. By increase of quartz it becomes micaceous quartzite or quartz schist, and by diminution of its crystalline character it passes into Argillite.

Hornblende Schist is a laminated mixture of hornblende with quartz, and sometimes with mica.

Talc Schist is a slaty rock in which Talc takes the place of Mica.

Chlorite Schist is a similar slaty rock consisting largely of that mineral.

Nacreous or *Hydro-mica Schist* is a name which has been given to crystalline slates in which a hydrous Mica takes the place of the ordinary Mica.

Marble or *Crystalline Limestone* and *Crystalline Dolomite* include the varieties of these rocks in which a perfect crystallization and often a white colour have been developed by metamorphism. *Ophiolite* is a marble containing grains or streaks and patches of Serpentine.

Anthracite and *Graphite* result from the alteration of Coal or of bituminous matter. Thus ordinary coal passes, under alteration, into anthracite, and finally, in certain cases, into graphite, and bituminous shales pass into graphitic slates.

Magnetite is very often a product of the metamorphism of ores consisting of the sesquioxide of iron.

Local metamorphism can often be observed at the contact of aqueous rocks with the larger igneous masses, and a study of these cases affords a key to the explanation of those larger examples in which no obvious cause of alteration is present. Metamorphism is induced or favoured by heat, by pressure, and by the percolation of heated and mineral waters; and rocks of complex character and containing basic and acidic minerals intermixed are those which present the most remarkable metamorphic changes. Such rocks have abounded more especially in the oldest rock formations, and in those partly made up of igneous ejections. At the same time the older deposits and those near to igneous foci have been the

most exposed to metamorphic agencies.   Hence certain metamorphic or
crystalline rocks are characteristic of the older formations, though not
absolutely confined to them.

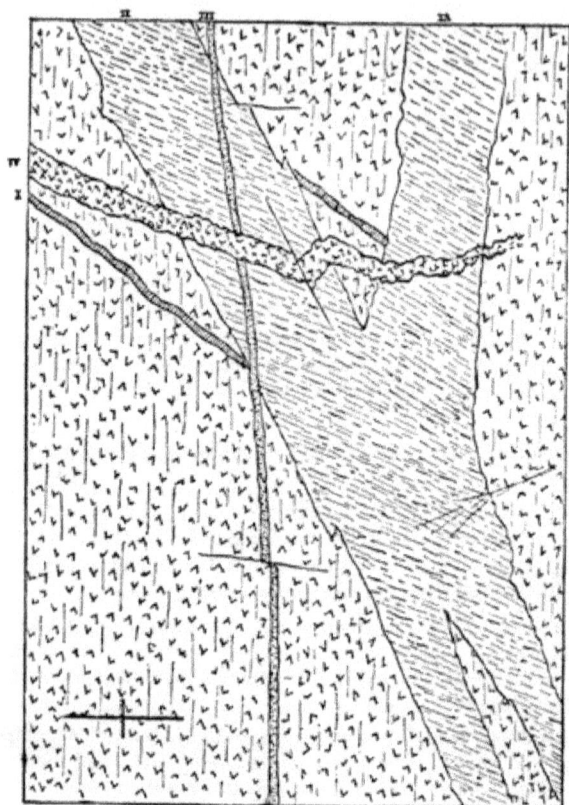

Fig. 1.

Metamorphic rock (Gneiss) intersected by Igneous dykes.   Lake of the Woods.   (I)
Red Felspar. (II) Greenish Diorite. (III) Hornblendic Diorite. (IV) Red Granite.
(G. M. Dawson.)

Scale—6 feet to an inch.

The conclusions of geologists have from time to time varied greatly as
to the causes and extent of Metamorphism of rocks.   These differences
of opinion have, however, like many similar disputes, been to some extent
subjective rather than objective, and have depended on the capacity of
observers to comprehend the phenomena which they have studied.

As to the facts, the conversion of woody matter into Anthracite, and
Graphite, and finally into Diamond, the change of ordinary organic lime-

stones into crystalline limestones, with various disseminated minerals contained in them, the change of sand into quartzite, of clay into micaceous schists, and many similar metamorphoses are so common and well known that they cannot be disputed. Such changes may refer either to crystallization of rocks not previously crystalline, to recombination of the ingredients of originally elastic or organic rocks, or to the introduction of new mineral substances by water or in vapour, and the consequent development of disseminated minerals whose materials were not previously present. The first of these is what Bonney has called Metastasis, or changes of position of molecules. The others come under the name Metacrasis, or changes of combination. What has been called Methylosis, or change of substance, is altogether exceptional, and not to be credited, except on the best evidence, or in cases where volatile matters have been expelled, as in the change of hæmatite into magnetite, or of bituminous coal into anthracite.

As to the causes of such metamorphic effects they are to be sought in the action of internal heat, aided by heated waters, along with enormous pressure either vertical or lateral. Local examples show the efficacy of these causes on a limited scale, and enable them to be applied to the larger areas of metamorphosed rocks, or to the phenomena of what has been called *Regional* as distinguished from *Local* Metamorphism. In connection with this, it is to be observed that the facts already stated in connection with igneous rocks show that the lower and older portions of the stratified crust must have been subjected for long periods to the contact of heated magmas in the under-crust, and to the action of heated water and the mineral matter contained therein under intense pressure. We shall also find farther on that the earlier stratified rocks are of small relative thickness, and have been crumpled and folded by lateral pressure in such a manner as to produce changes of their texture, and also to subject them to the action of heat and heated waters under different conditions from those which would have applied to them when horizontal.

The study of metamorphic rocks, whether in hand specimens or under the microscope, involves a combination of the questions and methods already referred to under igneous and unaltered aqueous rocks; and this complexity causes it to be a subject of greater difficulty. A glance at the geological map, or at any of the general sections of Canadian rocks in this volume, will show that aqueous rocks, whether unaltered or metamorphic, largely predominate over the surface of Canada as in all the larger areas of our continents.

Slicing Rocks and Fossils for the Microscope.

The aid of the microscope is invaluable in examining rocks of most kinds, and especially those that are crystalline. They may be studied as opaque objects or in thin flakes broken off with a hammer, but much better in thin slices prepared by a skilful lapidary, or which the student may prepare for himself as follows :—

Slicing and polishing machines may be purchased in London or Berlin, or a lathe may be converted into a Lapidary's wheel.*

If not provided with such a machine, the lithologist may use the following method:—

A cast iron plate should be procured about 9 inches square ; two pieces of glass of the same size ; a Water-of-Ayr stone; coarse and fine emery and putty powder ; glass slides, covering glass and Canada balsam for mounting the specimens.

Chips are then broken off the rock to be examined with a small hammer. They should be about half an inch in diameter, and as thin as possible. One side of a chip is ground flat on an iron plate with emery and water. To facilitate this the chip may, if desired, be cemented with a cement made of resin and bees' wax to a piece of wood. When one side has been ground flat, the chip is washed and the surface is then polished on the glass plate with flour of emery, or on the Water-of-Ayr stone. When quite polished the chip is again washed, and its flat side is now cemented with hard Canada balsam, or with melted lac to a strip of glass, and the other side is ground till the chip becomes transparent or translucent, when it can be washed and examined with the microscope when moist and covered with a thin glass. If not sufficiently thin, it can be farther reduced with putty powder on glass, or on the Water-of-Ayr stone. Care is required in the last stage lest the chip be altogether ground away. Practice will soon give the required nicety of touch, even if there be some failures at first. It will be best to begin with rocks not too hard, brittle or opaque.

The chip having been reduced to a transparent film, may be warmed and covered with balsam, or with balsam dissolved in ether, and a thin glass cover applied to it and a label gummed on and marked. If, however, it is desired to have neat slides, the balsam may be softened by heat and the thin film of rock slipped off to a clean slide on which it may be mounted in balsam.

To determine the minerals in the rock, the microscope should be furnished with a polariscope of two Nichol's prisms. For special lithological research, microscopes furnished with many appliances are now made ; but for ordinary purposes any good working microscope furnished with a polariscope will suffice.†

The above directions apply equally to fossils, except that of these it is usually desirable to have two or three slices in different directions. In the case of fossil woods, for example, we require a transverse slice and two longitudinal, one radial and one tangential.

---

* Cotton & Jordan, Grafton St., Soho ; Cottell, 52 New Compton St., Soho : Fuess, Berlin. The latter makes small and cheap machinery. Cutting Discs are made by Ken, 35 Westminster street, Providence, R. I.

† The best books on Microscopic Lithology are those of Zirkel and Rosenbusch, (German); and Rutley, "The Study of Rocks."

II. PALÆONTOLOGY.

(1.) Preservation of Organic Remains.

This depends in the first instance on the accidental imbedding of animals and plants or of portions of them, in deposits in process of formation, or on the accumulation of the remains of animals and plants on the surfaces on which they live, as for example of shells and corals on the sea bottom, or of vegetable matter in bogs and swamps. In one or other of these ways most aqueous deposits become more or less charged with organic remains. These are sometimes entire and sometimes fragmentary, and as already stated some beds contain so great abundance of organic fragments that they may be regarded as organic rocks. Often, however, the presence of organic fragments can be detected only by the lens or the microscope. (Figs. 2, 3.)

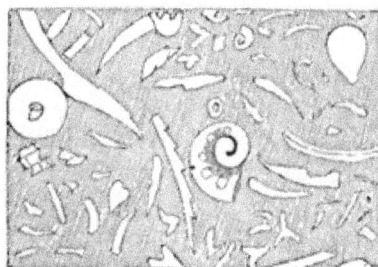

Fig. 2.—Fine grained Trenton limestone, Montreal, showing organic fragments × 10.

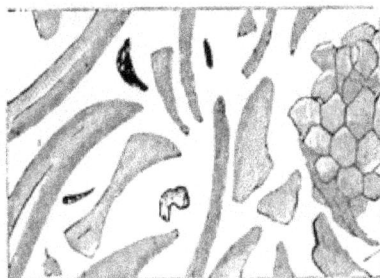

Fig. 3.—Chazy Limestone, Island of Montreal, showing fragments of shells and Stenopora × 10.

Organic remains may occur in an unchanged condition or only more or less altered by decay. This is often the case with such enduring substances as shells, corals, bones and wood, especially in the more recent

B

deposits, in which such remains occur little modified or perhaps only slightly changed by partial decay of their more perishable parts, as, for instance, of the animal matter of bones. In the older formations, how-ever, organic remains are usually found in a more or less mineralized con-dition, in which their original substance has been wholly or in part replaced by mineral matter, or has been chemically changed. The more important of these changes are the following :—

(a) *Infiltration* of mineral matter which has penetrated the pores of the fossil in a state of solution. Thus the pores of fossil wood are often filled with calcite, quartz, oxide of iron or sulphide of iron, while the woody walls of the cells and vessels remain in a carbonized state. (Fig. 4.) Bones, shells and corals in like manner have their cavities filled with mineral matter, and are rendered hard and heavy thereby. In the sea bottom the filling material is not infrequently composed of Glauconite or other hydrous silicates. (Fig. 5.) We sometimes find on microscopic examination that even cavities so small as those of vegetable cells and vessels have been filled with successive coats of different kinds of mineral matter.

Fig. 4.—Discigerus tissue (a, b), and Scalariform tissue (c) from carbonized plants of the Devonian system, highly magnified.

(b) Organic matters may be entirely *replaced* by mineral substances. In this case the cavities and pores have been first filled, and then, the walls or solid parts being removed by decay or solution, mineral matter either similar to that filling the cavities or differing in colour or com-position, has been introduced. Silicified wood and silicified corals often occur in this condition. In the case of corals and similar calcareous structures included in limestone, it sometimes happens that the walls of the corals are silicified while the cells are filled with limestone. Fossils thus preserved often appear with great distinctness projecting from the weathered surfaces of the containing rock. (Fig. 6.) In the case of silicified wood, it sometimes happens that the cavities of the fibres have been filled with silica and the wood has been afterwards removed by decay, leaving the casts of the tubular fibres as a loose filamentous substance. The more important of the foregoing modes of preservation are represented in Fig. 7.

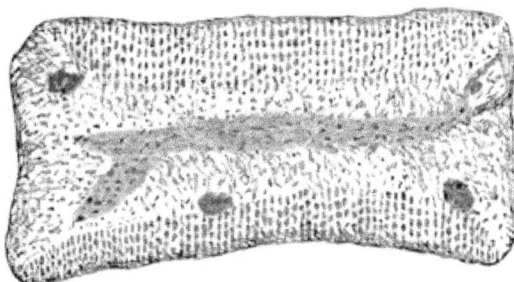

Fig. 5.—Joint of a Crinoid having the pores filled with a hydrous silicate allied to Glauconite. Upper Silurian, New Brunswick. Magnified.

Fig. 6.

Fig. 6.—Silicified corals, *Petraia pygmea*, and crinoidal joints weathered out on a limestone surface. (After Billings.)

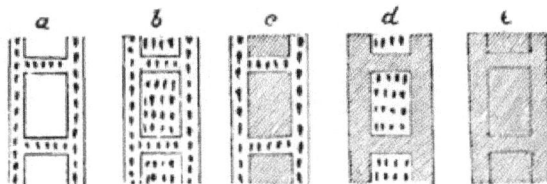

Fig. 7.—Sections of part of a cell of a Tabulate Coral in different states of preservation.

(a) Cell-wall calcite, cavity empty.
(b) Cell-wall calcite, cavity filled with the same.
(c) Cell-wall calcite, cavity filled with silica or a silicate.
(d) Cell-wall replaced by silica, cavity filled with calcite.
(e) Cell-wall replaced by silica, cavity filled with silica.

(c) The substance of organic remains may be wholly removed, leaving mere *moulds* or *impressions* of their external forms, or perhaps moulds of the external forms and casts of the interiors. This frequently occurs on the surfaces of rocks, where for example calcareous fossils have been weathered out from a harder matrix, but it also occurs in the interior of porous beds, owing to the solution of the fossils by percolating waters. In the case of fossils in this state, it is always necessary to consider whether the impression observed is that of the true exterior surface, of an inner layer, or of an interior cavity.

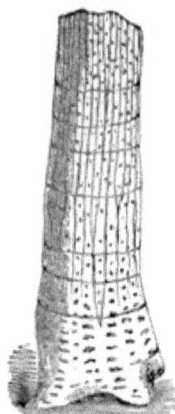

(d) The cavities left by fossils which have decayed may be filled with clay, sand or other foreign matter, and this becoming subsequently hardened into stone may constitute a *cast* of the fossils. Trunks of trees, roots, &c , are often preserved in this way, appearing as stony casts, often with the outer bark of the plant forming a carbonaceous coating on their surfaces. (Fig. 8.)

Fig. 8. - Trunk of Sigillaria represented by a sandstone cast of the interior of the bark. Coal formation of Nova Scotia. Reduced.

Fig. 8.

Fossils preserved in the two first modes usually show more or less of their minute structures under the microscope. These may be observed, (1) By breaking off small splinters or flakes and examining them either as opaque or as transparent objects. (2) By treating the material with acids, so as to dissolve out the mineral matters or portions of them. This method is applicable to some fossil woods, silicified corals, &c. (3) By grinding thin sections. These are first polished on one face, then attached to glass slips by a transparent cement or Canada balsam, and ground until they become so thin as to be translucent. (See p. 16.)

In the movements to which rocks have been subjected, fossils have often been distorted by pressure, whether vertical or lateral. Thus trunks of trees originally round have been flattened by the vertical compression of the beds, and where lateral pressure has affected the containing rocks, shells and other fossils have been shortened, lengthened or distorted obliquely. (Fig. 9.)

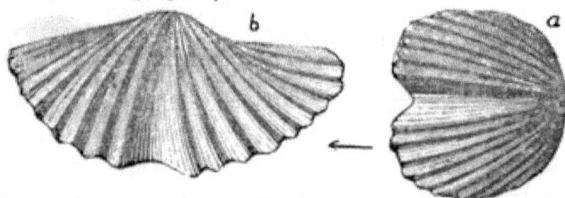

Fig. 9.—*Spirifer Nictarensis*, Lower Erian of Nova Scotia. *(a)* Shortened, and *(b)* lengthened, by distortion, in the direction of the arrow. (Acadian Geology.)

Ichnites or fossil footprints and similar markings constitute a peculiar and sometimes interesting kind of fossils. Animals walking over muddy shores may leave impressions, which being partially hardened by the air and sun, may not be obliterated by the succeeding deposits of sand or mud. Once so covered up, they remain for an indefinite time, and if the beds be hardened into stone, the footprints appear distinctly as the layers are removed by the quarrymen. In this way the footprints of some land animals, not known to us by other remains, have been preserved, and important information has been obtained as to their affinities and habits. (Fig. 10.)

Fig. 10.—Footprints of a Batrachian (*Sauropus*,) in ripple marked Sandstone. Coal field of Cape Breton.

Not only land animals, but aquatic creatures, as fishes, crustaceans, worms and mollusks, have left impressions and trails on the surfaces of beds, and these though less definite than the footprints of land animals, are of some importance as fossils. Such impressions have sometimes been mistaken for fossil plants ; but they can be distinguished by the absence of carbonaceous matter, by their close connection with the substance of the containing beds, by their being in relief on the under side of the beds, and by their forms. (Fig. 11.)

The geological observer in examining any section or exposure of rocks, while noting all the facts respecting stratigraphical arrangement and

relations, carefully collects the fossils of each bed, and labels them in such
a manner that their order of succession can be preserved.   The study of
these fossils may be expected to afford important information respecting
the age and conditions of deposition of the beds.   Should the observer
not possess the special knowledge necessary to interpret the fossils
obtained, he has recourse to palæontological specialists, either experienced
in the fossils of the formation in question, or of the groups of animals or
plants represented in the collections.

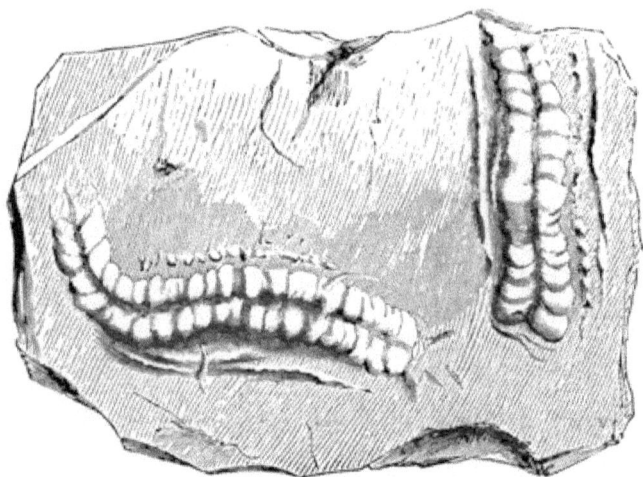

Fig. 11.—Tracks probably of a Crustacean (*Rusichnites*) in relief on under side of slab.
Coal formation of Cape Breton.

The most abundant and characteristic fossils available to the
palæontologist are those of aquatic animals, having hard shells, crusts or
cells.   Thus practically the most important elementary knowledge of the
study of fossils is that relating to the characters of invertebrate animals,
and especially those of the sea.   The student should therefore have some
familiarity with this subject, and should have for reference some good
zoological text-book, and if possible some work on the special palæontology
of the district or formation he may be studying.

In some geological formations, especially the middle and newer members
of the geological series, a knowledge of vertebrate animals becomes
important ; while in others, as the coal-formation, an acquaintance with
fossil plants is necessary.

(2.) CLASSES OF ANIMALS MOST IMPORTANT IN PALÆONTOLOGY.

The following table indicates the groups of animals most important to be known in connection with the study of fossils:—

PROTOZOA.

1. Rhizopods....................Foramanifera Polycistina.
2. Porifera.....................Sponges.

COELENTERATA.

3. Hydrozoa.....................Graptolites, Sertularians, &c.
4. Anthozoa.....................Coral Animals.

ECHINODERMATA.

5. Crinoidea ....Crinoids, or Feather-stars.
6. Asteroidea...................Star-fishes.
7. Echinoidea..................Sea-urchins.
8. Holothuridea ....Sea-slugs, &c.

MOLLUSCA.

9. Polyzoa ....Sea mats, &c.
10. Brachiopoda ....Lamp-shells, &c.
11. Lamellibranchiata............Ordinary bivalves.
12. Gasteropoda.................Snails and their allies.
13. Cephalopoda ....Nautili, Cuttlefishes, &c.

ANNULATA.

14. Annelida..................Worms.

ARTHROPODA.

15. Crustacea ....Soft shell fishes, &c.
16. Insecta ....Insects proper and Myriapods.
17. Arachnida...................Spiders and Scorpions.

VERTEBRATA.

18. Pisces ....Fishes.
19. Batrachia or Amphibia.........Frogs, Newts, &c.
20. Reptilia.....................Reptiles proper.
21. Aves.........................Birds.
22. Mammalia...................Mammals.

(3.) CLASSIFICATION OF PLANTS FOR PURPOSES OF PALÆONTOLOGY.

CRYPTOGAMS.

Thallogens, or
Thallophyta ....
- Algæ .... Sea weeds.
- Lichenes .... Lichens, &c.
- Fungi .... Mushrooms, &c.

Anogens, or
Bryophyta. ....
- Musci. .... Mosses.
- Hepaticæ .... Liverworts.

Acrogens, or
Pteridophyta ...
- Filices .... Ferns.
- Lycopodiaceæ. .... Club-mosses.
- Equisetaceæ .... Mares-tails.

PHAENOGAMS.

Gymnosperms, or
Archispermae ...
- Cycadeæ. .... Cycads.
- Coniferæ .... Pines, &c.

Angiosperms:

Endogens .... Numerous families of palms, grasses, and allied plants, with Monocotyledonous embryo.

Exogens . .... Numerous families of exogenous and covered-seeded plants, with Dicotyledonous embryo.

For further information as to fossil animals and plants, and more especially as to their characters and classification, the student may be referred to the author's "Hand-book of Zoology " and "Geological History of Plants," and also to such larger works as Nicholson's " Palæontology" and Zittell's " Palæontologie." With reference to the special Palæontology of Canada, reference may be made to the various Memoirs of Billings, Whiteaves and the author, in the Reports of the Geological Survey of Canada, and to the Palæontology of New York, by Dr. James Hall. Illustrations of some characteristic fossils will also be found in the following pages.—

* To these may be added the extinct group of *Protogens* having Algoid structure but constituting forests in early Palaeozoic times.

III. PHYSICAL GEOLOGY AND STRATIGRAPHY.

The previous departments relate principally to the study of hand specimens. This introduces us to the arrangement of rocks on the large scale, or to the manner in which they are built up as constituents of the crust of the earth. As the greater part of rocks are bedded or stratified, and the massive and vein-formed rocks may be considered as subordinate to the bedded in all those parts of the crust accessible to us, the term Stratigraphy may be used to cover the whole subject of *rock arrangement* on the large scale.

(1.) ORIGIN OF AQUEOUS DEPOSITS.

It will be useful in introducing this subject to notice in the first place the mode of formation of stratified rocks, and other matters connected with their present condition and appearance.—

In nature there is a constant struggle between aqueous and igneous agencies in modifying the materials of the earth's crust. The deeper portions of the crust are being slowly softened and crystallized under the influence of heat and pressure, and are thus being converted into metamorphic rocks, and these finally into plutonic masses, portions of which being erupted constitute volcanic products. On the other hand the waters and the atmosphere are constantly decomposing and wearing away the crystalline rocks at the surface, and depositing their detritus in the bottom of the waters. These processes seem to have been active, throughout the whole of geological time, in producing igneous and aqueous rocks. Since, however, the latter are the more important in geology, on account of their greater relative abundance, their regularly bedded character and the fossils they contain, we may direct our attention in this place principally to them, having already shortly noticed igneous phenomena under a previous head.

Atmospheric Erosion.—We have seen that the most common crystalline rocks are composed largely of silicates, as the Felspars, Hornblende and Pyroxene. When these are exposed to the action of the atmosphere and of rain water, which always holds carbon dioxide in solution, the soda, potash, lime, and other bases which they contain in combination with silica, are gradually removed in the state of carbonates, leaving the alumina and silica behind in an incoherent state. Thus from the decay of a hornblende granite there may result quartz-sand, clay, limestone, and iron oxides, which, when sorted and variously deposited by water, may assume the appearance of distinct alternating beds, while the alkaline

matters removed in solution are washed into the sea or into lakes, where
they may aid in chemical changes leading to other kinds of deposition.

To the atmospheric agencies we may also add the disintegrating power
of frost, which, by the expansion in the act of freezing of the water contained in rocks, chips off sand and fragments, and rapidly reduces very
hard rocks to ruins. In mountains and the polar regions this action of
frost is aided by the mechanical movement of glaciers, which removes to
lower levels or into the sea the material disintegrated by frost, and which
also exercises a polishing and abrading effect on the subjacent surface.
The action of coast ice, which is also very powerful, may rather be classed
with aqueous agencies.

Aqueous Erosion.—This takes place by the abrading action of rivers and
torrents, by the beating of the waves on coasts, by tidal currents, by the
action of cold heavy currents on the sea bottom, and by the solvent action
of springs and other subterranean waters. As these agents are constantly
at work, the changes which they produce in the lapse of ages are very great.
It has been estimated that the atmospheric and aqueous causes of erosion
at present in action, would suffice to remove the whole of the dry land
into the sea in about six millions of years.

Deposition.—The materials thus set free by chemical decomposition
and mechanical abrasion are deposited in layers in the depressed portions
of the earth's crust occupied by the waters. The coarser materials, as
pebbles and sand, may be thrown down along coasts and at the mouths of
rivers; the finer materials will be carried farther out to sea, and those
held in solution may be ultimately fixed in the organisms of coral animals
and other marine creatures, and may form coral limestones and similar
organic deposits.

In any given locality all these agencies, whether of erosion or of
deposition, may be greatly modified from time to time by changes of level
or of climate, whether arising from movements of the earth's crust, or
from astronomical causes; and also by volcanic paroxysms breaking forth
from time to time.

(2.) Hardening and Alteration of Aqueous Deposits.

Aqueous deposits thrown down by crystallization may be hard from the
first; but sedimentary beds are usually at first soft, and are hardened by
subsequent processes such as the following:—

(a) *By pressure* of a great thickness of superincumbent material. In
this way, for example, soft clay is hardened into shale, and peat into brown
coal; and there is reason to believe that lateral pressure, occasioned by

folding and settlement of the earth's crust, may produce still more powerful effects in hardening and crystallizing rocks. Pressure may act by con-
densing soft sediments to a fraction of their original thickness, by arranging flat particles on the same plane, thus causing lamination or cleavage,
by causing minute particles to adhere by contact, and by developing heat.

(b) *By Infiltration* of mineral matter in solution. Subterranean
waters usually contain calcium bicarbonate, soluble silicates or other
mineral substances in solution, and depositing these in the interstices of
sand, gravel, fragments of shells, &c., may ultimately cement such materials
into a compact rock. (Fig. 12.)

Fig. 12.—Fragment of Trenton Limestone, magnified. It is composed of broken
pieces of corals, crinoids and shells cemented together by transparent calcite.

(c) *By Heat.* When sediments are buried at so great a depth that
they are acted on by the earth's internal heat, or when heat is developed
by the movement and crumpling of great masses of rock, or when
sediments are invaded by intrusive molten rocks, they become baked and
hardened, and in some cases their particles are enabled to arrange themselves as distinct crystalline minerals or to enter into new chemical combinations. The result is metamorphism, which, as already stated, may
change mud or volcanic ashes or similar incoherent material into the
hardest and most crystalline rock. It is farther to be observed that the
heat to which sediments are subjected at great depths, is not dry heat,

since their substance is saturated with water, and this being prevented by pressure from escaping, remains in a heated state, and must greatly promote chemical and molecular changes.

(3.) CONCRETIONARY ACTION.

An important modification of these hardening processes results from concretionary action. This is an unequal hardening of the mass, whereby certain portions of it become indurated into balls, nodules or grains. It depends upon molecular attractive movements collecting together certain constituents of the mass, and may produce the following kinds of concretionary structure :—

(a) The whole mass of material may assume a concretionary structure, aggregating itself into nodular grains. This is the case with *Oolitic lime-stones* and *Oolitic ores of iron.* (Fig. 13.) A similar change sometimes occurs in the cooling of igneous masses.

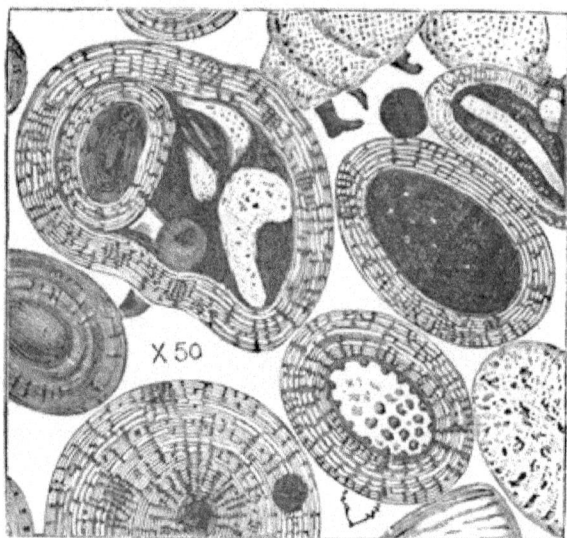

Fig. 13.—Magnified section of Oolitic Limestone (after Sorby,) showing concretions with radiating and concentric structure, and some of them enclosing fragments of shells, &c.

(b) Foreign materials diffused through the mass may be collected into limited spaces, and thus form concretions. This is the case with *flints* in chalk and with *clay ironstone* in beds of shale.

(c) The cementing substance of the mass may be unequally collected in certain portions at the expense of the rest. This occurs in the hard concretions in clays and in " bull's-eyes " in sandstone.

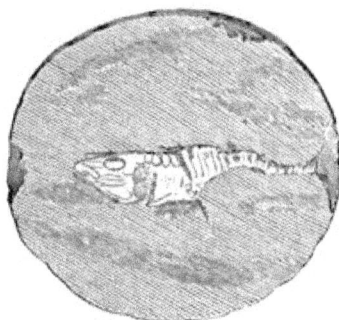

Fig. 14.—Rounded concretion containing a fossil fish, (Gasterosteus,) split open. Pleistocene, Canada.

Any foreign body, as a fossil or a grain of sand, may form a nucleus for a concretion. (Fig. 14.) Concretions have often a concentric lamination marking their stages of increase. They are sometimes hardened at the surface while the interior remains soft, and the latter may subsequently crack from shrinkage. When these cracks are afterwards filled with other mineral matter, *Septaria* concretions result. Concretions often assume very fantastic shapes, and have been mistaken for fossils.

The curious conical bodies found abundantly in some beds of shale and greenstone and known as "*Cone-in-cone*," (Fig. 15,) appear to be caused by concretionary action proceeding from the surface of a bed or layer and modified by the gradual compression of the material.* A similar origin has been attributed to the columnar striated bodies found in some limestones and named *Stylolites*, which are often occasioned by the presence of shells or other objects on a soft mass undergoing compression.

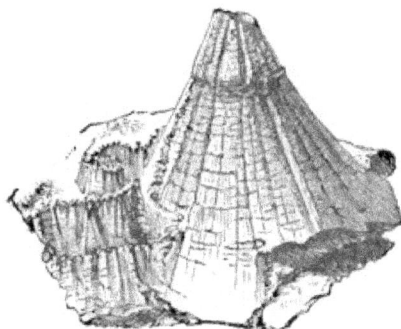

Fig. 15.—Cone-in-cone. Coal formation of Nova Scotia.

* See Acadian Geology, p. 676 for more full explanations.

Geodes, which are cavities in rocks lined with crystals, are distinct in their mode of formation from concretions, though sometimes confounded with them.

(4.) COLOURS OF AQUEOUS ROCKS.

The most abundant colouring matter in rocks is iron. Its monoxide and sulphide when diffused through sediments produce green, gray and blackish colours. Its sesquioxide produces red colours. Its hydrous sesquioxide gives yellow, buff and brown shades. Peroxide of manganese is sometimes a cause of black colours in rocks, and coaly matter is also a not infrequent cause of the blackening of sediments.

The following facts are important with reference to the colours produced by iron :—

(a) In the subaerial decomposition of most rocks a sufficient quantity of sesquioxide of iron is produced to colour the resulting sands or clays. In ordinary circumstances it is the brown or hydrous oxide that is produced in this way ; but in warm climates, under the influence of volcanic heat and in the presence of saline waters, the red oxide is produced. Thus the subaerial decomposition of crystalline rocks coloured gray, green or black by sulphide or monoxide of iron, gives rise to brown and red sediments.

(b) If the sediments thus coloured are rapidly washed down and deposited in the sea, or in limited areas of fresh or salt water, they may retain their colours, and thus the red, brown and purple sandstones and clays so characteristic of certain formations are produced.

(c) If the sediment is long abraded by moving water, the clay is separated from the sand, and the superficial red coating is washed from the latter so that it loses its colour. In this way gray or white sandstones are often found to alternate with red or reddish shales.

(d) When sediments coloured with iron are deposited in fresh water along with organic matter, as peat, &c., the latter deprives the iron of a portion of its oxygen, reducing it to monoxide, and this being soluble in the acids naturally produced by the decay of the vegetable matter, is removed, leaving the sand or clay in a bleached condition.

(e) When the deoxidizing process occurs in sea water, the sulphates present in the latter being decomposed at the same time with the iron oxides, a black iron sulphide is produced, which gives a gray colour more or less dark to the sediment. Material coloured in this way becomes buff or brown on weathering, and becomes red when heated in the air. This is a useful mark of marine clays. In this case or the last, scattered organic fragments deposited in red sediments and not in sufficient quantity to affect the colour of the whole, produce gray or white stains.

(f) If organic matter be present in large quantity, it not only removes the red colour but communicates its own black or dark brown colours to the whole.

The above considerations serve to show why red rocks have been deposited in large quantity in times of physical disturbance and volcanic activity, and generally when deposition is rapid and organic matter absent. They also serve to explain the presence of red beds with rock salt deposited from the waters of saline lakes or lagoons. They also explain the rarity of fossils in red rocks, since the retaining of the red colour implies scarcity of organic remains, and an excess of peroxide of iron tends to oxidise and destroy such as may be present. On the other hand they show why gray and dark coloured beds are those which most abound in fossils.

(5.) MARKINGS ON THE SURFACES OF AQUEOUS ROCKS.

The circumstances under which aqueous beds have been deposited are often indicated by markings seen on their surfaces.

(a) *Ripple marks*, caused by the motion of currents throwing up slight ridges and hollows at right angles to the direction of the current.

(b) *Current lines*, caused by the driftage of sand, organic fragments, or sea-weeds and drift wood, in the direction of the current.

(c) *Rill marks*, caused by the running of drainage water over inclined surfaces of mud and clay after recession of the tide. These are often so complicated as to simulate foliage and have frequently been mistaken for fossil marine plants. (Fig. 16.)

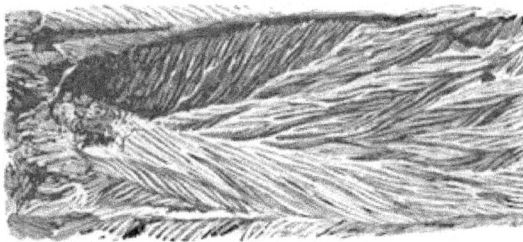

Fig. 16.—Rill-marks, Carboniferous (reduced.)

(d) *Shrinkage cracks*, produced by the drying and shrinkage of muddy surfaces when left bare to be acted on by the sun and air. (Fig. 17.)

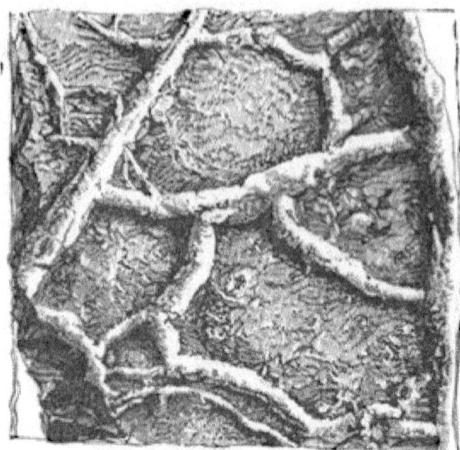

Fig. 17.—Shrinkage cracks, Carboniferous (reduced.)

(e) *Rain marks*, or rounded pits produced by rain drops, or washed surfaces produced by continuous rain, afterward covered up and preserved by subsequent deposits. (Fig. 18.)

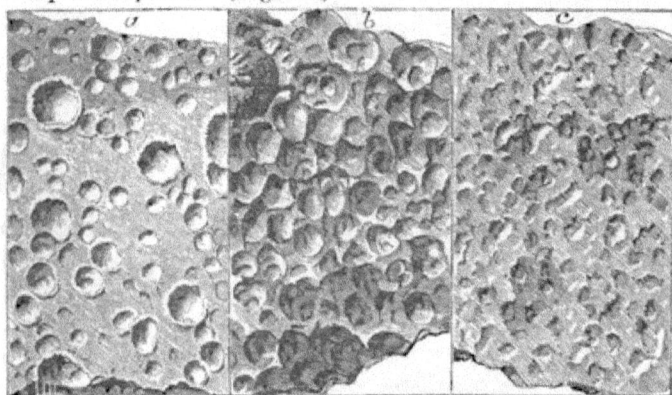

Fig. 18.—Rain-marks, *(a)* modern, *(b c)* Carboniferous.

These markings belong for the most part to shallow water and to the vicinity of the shore and to tidal estuaries. They are often of much interest as indicating the conditions of deposit and the changes which have taken place in these.

(6.) Arrangement of Rocks on the Large Scale.

With reference to this, the materials of the earth's crust exist in three different conditions :—(1) *The Stratified ;* (2) *The Massive or Unstratified ;* (3) *The Vein-formed.* The rocks of the second and third classes are however subordinate to those of the first, which vastly predominate in those parts of the earth open to our inspection. We may therefore consider first and principally the Stratified rocks. (Figs. 19 and 20.)

All ordinary aqueous or sedimentary rocks are stratified, or arranged in beds more or less nearly, when undisturbed, approaching to a horizontal position.

A *Lamina* or *Layer* is the thinnest sheet into which a stratified rock is divisible. Some shaly beds are divisible into extremely thin laminæ. Other beds are destitute of lamination and are said to be compact.

A *Stratum* or *Bed* is of greater thickness, or may consist of several laminæ—*e. g.* a bed of laminated sandstone or shale consisting of several layers.

The term *Seam* is often used by miners for beds of useful minerals; and when such beds are considerably inclined, they are sometimes called veins, though not of the nature of true veins.

A *Formation* consists of several beds deposited consecutively and under similar general conditions. A formation may thus include beds of rock of different kinds, though usually there is a certain lithological similarity in the beds constituting a formation—*e. g.* the coal formation, which includes many beds of sandstone, shale, coal, &c., or the Laramie Series of the West. (Fig. 20.)

The idea conveyed by the term "Formation" is one of the most important in Geology, since it relates not to lithological similarity but to continuous deposition under like general conditions. The more important formations are designated by the term Series, and subdivisions of such formations by the word Stage. (Fr. *étage*.)

By French Geologists the word "Terrane" is used in a general sense to indicate any geological formation.

Fig. 19. — Section through Montreal mountain, showing massive igneous rock at (a); dykes at (dd); and bedded or stratified rocks at (b, e, f, g.) The lower beds of the latter belong to the Siluro-cambrian age, the upper (e, f, g) to the much later Pleistocene formation.

A System of Formations includes all the formations of one of the larger geological periods—*e. g.* the Carboniferous System, which includes with the coal formation other formations belonging to the same great geological period.

The term *Group* has been proposed for the larger Divisions embracing several systems ; but this term is more usually and properly employed to designate any assemblage of strata or formations.

Inasmuch as formations and systems of formations imply the lapse of time, they may also be designated by terms relating to time. Thus we may speak of the Carboniferous Period, the Coal-formation Epoch.

The classification of formations in relation to time will be considered under the heading of Historical Geology.

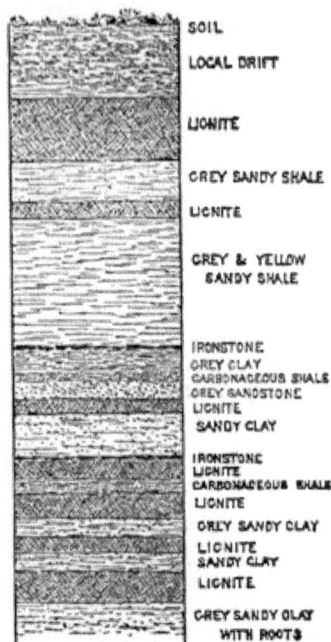

SOIL
LOCAL DRIFT
LIGNITE
GREY SANDY SHALE
LIGNITE
GREY & YELLOW SANDY SHALE
IRONSTONE
GREY CLAY
CARBONACEOUS SHALE
GREY SANDSTONE
LIGNITE
SANDY CLAY
IRONSTONE
LIGNITE
CARBONACEOUS SHALE
LIGNITE
GREY SANDY CLAY
LIGNITE
SANDY CLAY
LIGNITE
GREY SANDY CLAY WITH ROOTS

Fig. 20—Section of Laramie formation, west of Manitoba. The whole of the beds shown, except the soil and drift, belong to one Formation, though differing in mineral characters. Some of them, as the shale beds, are laminated. (G. M. Dawson.)

(7.) JOINTS AND SLATY CLEAVAGE.

These appearances are important, because it is necessary to distinguish them from planes of bedding.

Joints are planes of division cutting beds at various angles, though usually approaching to vertical. They often divide the bed into oblique angled blocks by the intersection of two sets of cleavage planes ; and when the cleavage planes of one set are close together they often simulate true bedding. Joints sometimes facilitate the operations of the quarry-man by enabling blocks of stone to be more readily detached ; but when numerous they injure stones otherwise useful.

When joints occur in beds of igneous rock they sometimes give origin to a *columnar structure*, as in beds of basalt.

Joints are often *slickensided*, that is, they have their surfaces polished by friction which has occurred during movements of the beds. Joints may sometimes have been produced by such mechanical movements; but are usually attributed to shrinkage or to a rough kind of crystalline cleavage.

Joints are sometimes widened into fissures, which being filled with foreign matter, constitute veins. The joints of rocks thus connect themselves with the vein-condition afterwards to be noticed.

Jointed structure sometimes weakens otherwise enduring rocks so as to permit them to be worn into ravines and valleys.

The student should observe here that in cases where intense pressure has been applied to massive igneous rocks they have sometimes assumed a laminated structure similar to that of metamorphosed aqueous beds. Such phenomena are usually local, and are to be distinguished by study of the relations and structure of the rocks concerned.

Slaty structure is a lamination not caused by original deposition but by pressure subsequently exercised, whereby plates of mica and other flat bodies present in the material, may be induced to assume positions parallel to the plane of pressure. Such slaty structure or slaty cleavage has been effected in many regions over great thicknesses of beds, and while it is of practical importance as giving the best roofing slates, it is somewhat puzzling to geologists as masking the true bedding. This can however be usually ascertained by noticing the bands of colour and structure which represent the original planes of deposit. In some cases the planes of bedding and of cleavage coincide, but in very many they are altogether different. (Fig. 21.)

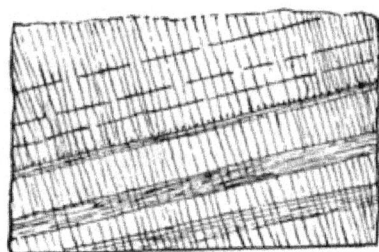

Fig. 21.—Silurian shales affected with slaty cleavage, Matapedia River. The bedding is represented by the slightly inclined lines; but the rock cleaves across the bedding in the planes indicated by the more highly inclined set of lines.

(8.) Inclined Position of Beds.

Aqueous strata have been originally deposited in a position approaching to horizontality. The only exceptions to this are where these beds have been uniformly disposed over uneven or inclined surfaces, or where material has been washed over the edge of a bank, giving rise to oblique stratification or false bedding. Movements of the crust of the earth, and especially movements of folding or bending which have apparently arisen from the shrinkage of the mass of the earth as compared with its crust, have however caused the originally horizontal beds to assume various degrees of inclination. Aqueous erosion has further caused the broken edges of bent strata to protrude at the surface. The degree and direction of such inclination afford most valuable data for ascertaining the relative positions and ages of beds. The most important facts of this kind are the following :— (Fig. 22.)

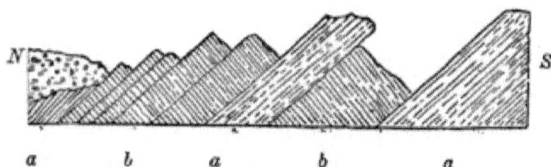

Fig. 22.—Inclined position of rocks. Beds of slate *(a,)* and iron ore *(b)* dipping to the northward at an angle of 41°. (Pictou, N.S.)

(a) *Dip*, or the angle of inclination to the horizon as measured by the clinometer.

(b) *Direction of Dip* as ascertained by the compass.

(c) *Strike*, or the horizontal line at right angles to the dip.

(d) *Outcrop*, or the line of intersection of the plane of the bed with the surface of the country. On perfectly level ground this is of course identical with the strike. Otherwise it is different.

Observations of these facts can be made in natural exposures, as cliffs and shores, or in artificial exposures, as quarries, cuttings, mines, &c. The harder rocks usually project in ridges and the softer are cut into hollows. Hence the lines of ridges and valleys often form very useful guides in tracing the outcrops of beds. The harder rocks are also more likely to crop out at the surface than those which are softer, and the latter are more liable to lie in low ground and to be covered with soil.

A line drawn across the strike of a series of beds gives a section of those beds, and in proceeding along such a line in the direction towards which the beds dip, we obtain an *ascending series*. In the opposite direction we obtain a *descending series*. Thus we can ascend or descend geologically in proceeding along the surface of the ground, and geological ascent and descent do not coincide with topographical, except where the beds are horizontal or nearly so, or when they dip toward the summit of an elevation.

The thickness of beds is always measured at right angles to their dip. For ordinary purposes it may be assumed that the thickness is equal to $\frac{1}{12}$ of the distance across the outcrop at 5° of inclination, and so on for every additional 5°.

When we follow a series of beds in ascending or descending order, we at length arrive at a line in which their dip changes to the opposite direction. When this takes place in the descending series it constitutes an *anticlinal line* or *axis*, sometimes called an *anticline*. When it takes place in the ascending series it constitutes a *synclinal line* or *syncline*.

When the anticlinal and synclinal axes are not horizontal, or when the surface of the country is inclined, the beds may be seen at the surface to bend around the ends of the anticlinals or synclinals, so that on a map these appear as more or less abrupt bends or loops of the strata.

In those regions where the beds have been slightly inclined, the anticlinals and synclinals are low and wide ; but in disturbed districts the folds are often very abrupt, causing the beds to approach to verticality, and in some places to be overturned. In such cases also the anticlinals or synclinals are sometimes very steep on one side and less so on the other, and they are not infrequently accompanied with minor flexures and foldings of the beds as well as with fractures or dislocations. In such disturbed districts great caution is requisite lest abruptly folded and repeated beds should be regarded as constituting a continuous series, and lest overturned beds should be regarded as in their natural positions. (Figs. 23, 24, 25.)

Fig. 23.—Anticlinal fold, Lignite Tertiary formation, showing also denudation of the axis of the anticlinal. St. Mary River, N. W. T. (After G. M. Dawson.)

Fig. 24.—Contorted Laurentian rocks, Ottawa River. (After Logan.) (a, b) Lower Laurentian Gneiss and Limestone. (c) Upper Laurentian. (d) Mass of Granite. (e) Dyke of Porphyry.

Fig. 25.—Beds of Limestone, Sandstone and Shale of Lower Carboniferous age in a vertical position. Smith's Island, Cape Breton.

(9.) FAULTS.

When movements of beds have been accompanied with fracture and slipping of the beds up or down, faulting or discontinuity of beds is produced.

Faults traversing inclined beds may displace them laterally as well as vertically. The vertical displacement is sometimes designated by the term slide, the lateral displacement by the term heave. A *downthrow* is said to take place on that side toward which the beds are sunken, and an *upthrow* on that side toward which they have risen. When the plane of a fault is inclined, the inclination is usually called by miners its "*hade*," and is measured from a vertical plane. The downthrow is almost always found to have occurred on the side toward which the plane of fault inclines. (Fig. 27.) When the contrary occurs the fault is said to be reversed, or inverted. (Fig 26.)

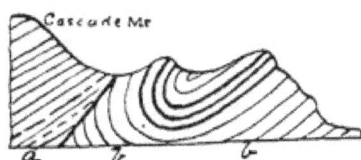

Fig. 26.—Reversed fault bringing Palæozoic rocks *(a)* over folded Cretaceous, *(b)* Cascade mountain, Rocky Mountains. (After McConnell.)

In highly disturbed districts reversed faults are not infrequent, and in some cases beds have even been thrust horizontally over others, producing very complex arrangements, and sometimes leading to error in estimating

the true ages of beds. Good examples of these abnormal arrangements
exist in the Canadian Rocky Mountains and in the highlands of Scotland.
(Fig. 37.) It often happens that several faults in the same direction
occur near each other, throwing the beds in steps, and in disturbed
districts there are often two or more sets of faults crossing the beds in
different directions. In this case the original position of the beds can
be ascertained only by careful study of the effects of the several dislocations. The surface of the plane of a fault is often polished and striated
or slickensided by the movement of the beds. Sometimes, however, the
edges of the beds have bent and broken where in contact, and sometimes
lines of fault present open fissures filled with debris or with crystallized
minerals.

In observing faults, the facts to be noticed are the directions of the
planes of fracture, their hade and the amount and direction of movement, with its effect on the beds traversed. When these facts are
obtained, all the effects of the dislocation can be readily worked out,
though, when several lines of fault cross the same beds, the appearances
are often very deceptive, leading to incorrect estimates of the thickness
and number of the beds.

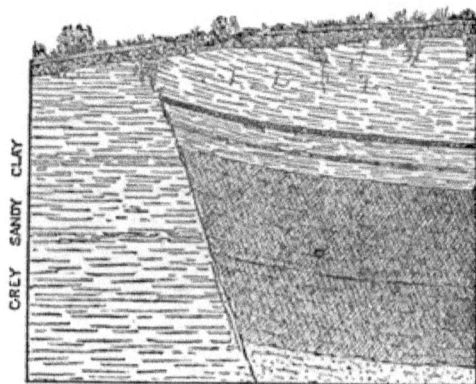

Fig. 27.—Fault, Lignite Tertiary series, Porcupine Creek, N. W. T.　(G. M. Dawson.)
The bed of lignite *(a)* has been thrown down, and has been removed by
denudation from the other side of the fault.

As the inequalities caused by faulting have usually been rounded or
smoothed off, and the line of a fault is often a weak place where the rocks
have been worn down and covered with debris, faults can very rarely be
distinctly seen, and their nature and direction can usually be ascertained

only by inference from the dislocation observed in the beds on their opposite sides. They are very numerous in disturbed districts, and there are often two or more sets of them crossing the beds in different directions. In most cases, however, the amount of movement which they produce is not great.

(10.) UNCONFORMABILITY.

When one series of beds has been disturbed and another deposited upon the upturned edges of the first, the upper series is said to rest uncomformably on the lower. This indicates not merely a difference of age but an interval of time between the dates of the two series. It often happens also that the edges of the lower series show evidences of great erosion, or that the beds of the lower series have been hardened and altered before the deposition of the upper. A false or simulated want of conformity occurs when a bed has been cut unequally by water before the next bed is deposited. When conglomerates or coarse sandstones rest upon finer beds such apparent unconformity is often produced. (Figs. 28, 29, 30.)

It is to be observed here that false bedding and also partial denudation may simulate true unconformity and that progressive subsidence may be accompanied by the overlapping of successive formations.

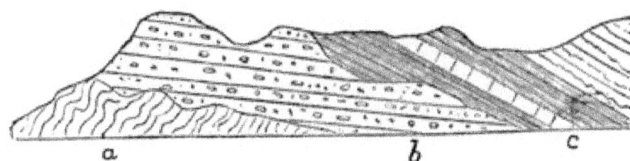

Fig. 28.—Unconformable superposition of *(c)* Silurian beds on *(b)* Cambrian, and of the latter on *(a)* Eozoic. West of Scotland. (After Murchison.)

(11.) DENUDATION.

This is the removal of matter by atmospheric or aqueous erosion. It has already been referred to as a source of the materials of aqueous deposits. We must now consider it as concerned in giving relief to the surface of the earth. That denudation has taken place to a great extent may be inferred from such facts as the following : The projection of hard beds and massive rocks in consequence of the removal of softer material from around them ; the existence of synclinal elevations in consequence of the erosion of anticlinals which once were higher but must have been more perishable owing to their fissured condition ; the planing away to a

low level of rocks which testify by their dips or by the existence of
extensive faults that they once rose to much greater height and were very
uneven ; the cutting of deep ravines through table lands, and the quantity
of stones, gravel, sand, and other detritus of older formations, employed
in the building up of those which are newer. (Fig. 31.)

Denudation of undisturbed beds also leaves exposed hard igneous
masses which may have penetrated them and which sometimes protect
portions of the surrounding beds partially hardened by contact with the
igneous rock.   Montreal Mountain and its companion eminences in the
valley of the St. Lawrence are apt illustrations.

In desert regions and along sandy coasts grains of sand carried by the
winds are capable of producing a certain amount of erosion.

Fig. 29.—Unconformable superposition of Triassic red sandstone (b) in highly
inclined carboniferous rocks, at Petit River, Walton, Nova Scotia.

Fig. 30.—Apparent unconformability produced by denudation and filling up. S.
Joggins, Nova Scotia. a, Sandstone. b, Shale. c, Irregularly bedded
Sandstone and Remains of Plants.

Geological observation has shown that the inequalities of the earth's surface are very largely due to denudation. This is best seen in sections

Fig. 31.—Denudation of horizontal beds, Great Valley, N. W. T.   (G. M. Dawson.)

of highly inclined rocks in which it often appears that such rocks have been planed off on the surface (Fig. 24,) while in other cases the softer rocks have been cut away and those of greater hardness remain as peaks or ridges.   (Figs. 19, 23, 25.)

It has been estimated that the areas drained by the rivers of our continents are losing by denudation at rates varying from 1 foot in 1500 years to 1 foot in 6000 years.   At these rates, were no counteracting elevation to take place, our continents would be levelled with the sea in from four millions to nine millions of years.

(12.) Massive Rocks.

These are in almost all cases of igneous origin, and can be readily distinguished from the stratified rocks both by their mineral character and their mode of occurrence.   Such irregular masses may represent either (1) the remains of the bases of old volcanic cones, the looser parts of which have been swept away ; or (2) exotic or intrusive materials ejected among other rocks from beneath ; or (3) portions of the aqueous crust so much altered that their stratification has been obliterated.

If the stratified rocks have been altered at their contact with igneous masses, or are penetrated by veins proceeding from them, we know that the masses are newer than the beds.   (Figs. 19, 32.)   On the other hand, if the massive rocks have been eroded before the deposition of the beds, if the latter are unaltered, and if they contain debris derived from the massive rocks, we know that these are older.

Fig. 32.—Junction of massive intrusive granite and Cambrian quartzite in cliff near Indian Harbour Lake, Nova Scotia.

(13.) VEIN-FORMED ROCKS.

The most common veins are fissures filled with material introduced either in a molten state or in aqueous solution.

Igneous veins or *dykes* are often of great size, and extend through the stratified rocks for long distances. They are filled with some of the kinds of igneous rock; sometimes present a jointed structure at right angles to their sides; often have the surface in contact with the adjacent rock of different texture from the interior, and have often, by their heat, produced considerable alteration in the adjacent rock. They are especially numerous in the vicinity of igneous masses and of volcanic foci ancient or modern. (Figs. 19 and 33.)

Veins or dykes, if harder than the enclosing rocks, may project above the surface like walls when the latter are removed by denudation. If softer than the containing rocks they may be eroded into ravines or furrows.

Aqueous veins, which are often *mineral veins*, are usually filled with crystalline minerals deposited in them by water. They often present a

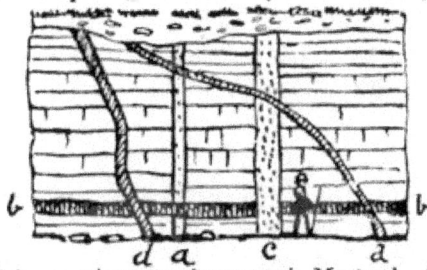

Fig. 33.—Igneous dykes or veins, extension reservoir, Montreal. *(a)* Felspathic dyke traversing beds of limestone. *(b)* Floor or horizontal vein of Dolerite cutting *(a)*. *(c)* Thick dyke of Felsite cutting *(b)*. *(d)* Inclined dykes of Dolerite cutting all the others.

laminated appearance, owing to the deposition of successive coats of matter on the walls of the vein. Occasionally the walls of veins present margins or "*selvages*" consisting of decomposed rock or decomposed veinstone.

In the case of mineral veins, the mass filling the vein is called gangue or veinstone, as distinguished from the ore associated with it. The rock traversed by veins is called by miners the "country rock."

Veins may fill open fissures, or may be of material slowly segregated into cracks in process of formation and of widening. They are usually vertical, or nearly so, but igneous veins have sometimes been injected between the beds, constituting "floors" or horizontal veins. (Fig 33.) Such horizontal veins when of considerable thickness have been named Laccoliths. Veins are sometimes lenticular or interrupted, and not infrequently occur at or near the junction of different formations. (Fig. 34.)

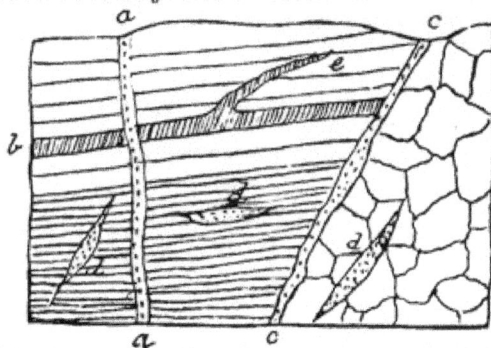

Fig. 34.—Metallic veins near the contact of slate and granite. (After Von Cotta.) *(a)* Fissure vein. *(b)* Horizontal or bedded vein sending off a branch *(c.)* *(c)* Contact vein at the junction of the two formations. *(d)* Lenticular or interrupted veins, sometimes called by miners "pockets."

Veins are often very irregular in their forms. This arises not only from the original irregularity of cracks traversing rocks, but from subsequent shifts of the containing walls, from detachment of loose pieces or "*horses*" from the sides, and from erosion of the walls by subterranean waters. They also differ very much in their contents in passing from one kind of rock into another, and are often decomposed and changed by atmospheric action at their outcrops.

There is reason to believe that some veins have been filled simultaneously with their opening, so that they have never actually been open fissures. Such veins, which present many peculiar appearances, are called segregation veins.

(14.) Origin of Mountains.

We are now in a position to consider the general principles which regulate Orography or the relief of the continents, as composed of Plains or Valleys, Table-lands, and Mountains or Mountain ranges.

(1.) Aqueous denudation may cut table-lands into deep ravines or *Canyons*, of which those of Colorado and of some districts of British Columbia are good illustrations. Where such cutting widens these ravines, it may produce flat-topped mountains, composed of horizontal or even of synclinal beds.

Fig. 35.—General Structure of the Cobequid Range, Nova Scotia. An anticlinal mountain, with unconformability and intrusive igneous masses and veins. a, Massive Syenitic Granite. b, Lower Cobequid Series, Felsite, Porphyry, Agglomerates, &c. c, Upper Cobequid Series. Ferriferous Slates and Quartzite. d, Wentworth Fossiliferous Beds.—Silurian. e, Carboniferous. f, Triassic. x, Veins of Syenite and Diabase.

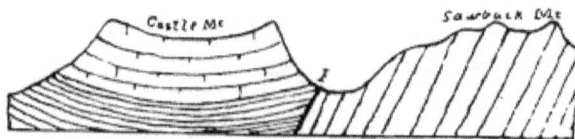

Fig. 36.—Synclinal and Monoclinal mountains, Rocky Mts. (after McConnell.) The beds are Cambrian, and are broken by a fault at E.

(2) When the earth's crust has been thrown into broad folds of small inclination, (geoclinal bends), these may form broad-topped elevations

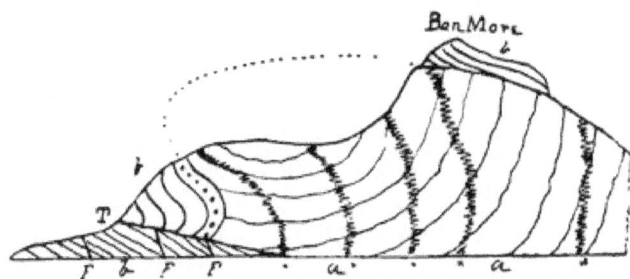

Fig. 37.—Ben More, Scotland, (after Geikie.) showing thrust (T) of contorted gneiss *(a)*, over Cambrian and Siluro-Cambrian rocks *(b)*, and denudation indicated by the dotted lines. Faults at *F F*. Dykes at x x.

subsequently cut by denuding agencies into mountain masses. The Lebanon range and some of the ridges of the Apalachians and in the western part of America furnish illustrations.

(3.) The abrupt crumpling and sliding of the crust of the earth is the cause of most of the larger and more important mountain ranges, though such ranges are always greatly modified by denudation.

(4.) In volcanic regions mountains of considerable elevation are often composed of the ashes, lapilli and lavas ejected by the volcanic vent, and arranged more or less in a conical form. Vesuvius is a heap of this kind 4,000 feet high, and which has been piled up since the year 79 of the Christian era.

In great mountain chains, as in the Cordillera of western America, all the above agencies may enter into the causes of mountain-making, and such great mountain ranges are often due to agencies which have been in operation either constantly or with intermissions throughout the whole of Geological time. It therefore usually happens that mountain ranges whose formation has been going on up to late geological periods are of larger dimensions and more complicated than those in which the elevatory movements have been arrested at early dates in the earth's history.

The experiments of Cadell and others, as well as the calculations of Prof. Darwin and Mellard Reade have shown that the thickness of the crust affected by these movements must be relatively very small, and I have shown that evidence of this fact is afforded by the great igneous dykes which traverse for long distances the crumpled Laurentian rocks of Canada, and cut directly through them as a thin veneer of curled wood may be cracked through by the shrinkage of the plank on which it is laid. Thus the wrinkled crust, including its mountain masses, is but a thin veneer on the homogeneous under-crust.

In some mountains great uplifts, fractures and lateral thrusts have given the original inequalities. In others the beds have been closely folded with little fracture. But while these causes have produced the original elevations, denudation has so greatly modified these that synclinal mountains are by no means uncommon. It is also to be observed that those belts along ocean margins where the greatest amount of deposition and downward bending of the crust have occurred, are often or usually those which have experienced the most violent folding.

(15.) CHRONOLOGY OF BEDS.

Superposition.—The great leading fact as to the ages of aqueous deposits is that the upper of two beds is necessarily the newer. Wherever therefore the actual superposition of beds can be ascertained, there can be no doubt as to their relative ages. In mines and borings, and in cliffs and quarries, we can thus easily ascertain the ages of the beds exposed. In the case of inclined beds this is equally obvious as in those which are horizontal. In these, however, we must be careful not to be misled by overturns and by repetition of beds by faults.

No one exposure, however, can show anything more than a limited portion of the series of rocks occurring in any district of considerable extent. Hence in extending the results of our observations it is necessary to have recourse to other data.

Tracing of beds.—Having ascertained the sequence of beds in one locality, we endeavour to trace them along their outcrops. We thus bring them into relation with other beds not seen in the original exposure.

Mineral character.—Where the tracing of the beds fails, we have to compare them in different sections, and to endeavour to recognize them by their mineral character—a succession of like beds in two not very distant sections giving us fair evidence of identity. Here, however, we must remember that in tracing any given bed for a long distance, it cannot be expected to retain precisely the same character, but may be represented by some different material.

Fossil remains.—When we can obtain from any of the beds in question fossil organic remains, these afford us a new means of testing identity of age. Experience has shown that in the course of the earth's history the facies of animal and vegetable life has been constantly changing, so that the fossils of one formation are different from those of another. When we have in any one locality ascertained this succession, we are safe in applying it to others. The evidence of fossils is thus at present held to be one of the best criteria for the ages of stratified rocks.

In employing fossils as evidence of age, we have, however, to bear in mind certain necessary precautions. There are other differences than those of age ; as for example, the difference between animals of the sea, of freshwater and of land ; of different depths in the sea, and of different climates. It is necessary, therefore, to compare animals or plants of like habitat and conditions of existence. It is farther to be observed that certain forms of life have been of longer duration in geological time than others, and therefore do not so definitely mark the lapse of time. Again, certain forms of animal or vegetable life may have been begun earlier or continued later in one locality than in others. On these accounts the evidence of fossils is more certain with reference to the greater geological periods than with reference to the minor subdivisions of these.

It has been supposed that the similarity of geological succession in distant places may not imply strict synchronism but merely what has been called *Homotoxis*, or resemblance of order without identity of time. When, however, we compare the whole series of stratified rocks in different regions we find that there is an identity of sequence throughout, from the earliest to the latest, so that there must be a general identity of time in the several members. We must, however, make allowance for the earlier or later introduction or extinction of the faunæ and floræ at the beginning and end of each great period.

Geological Cycles.—The foregoing considerations bring in a very distinct manner before us two different, and at first sight irreconcilable, general views which we may take of any given geological period. *First*, we must regard every such period as presenting during its whole continuance the diversified conditions of land and water with their appropriate inhabitants, or in detail—mountain ridges, continental plateaus and ocean depths ; and *secondly*, we must consider each such period as forming a geological cycle, in which such conditions to a certain extent were successive. As we give prominence to one or other of these views our conclusions as to the nature of geological chronology must vary in their character, and in order to arrive at a true picture of any given time it is necessary to have both before us in their due proportion.

We shall thus find that in each geological system there was at least one great continental depression bringing marine life over the continental plateaus, and this with elevations before and after. In the meantime the higher parts of the continents may have been continuously land, and the ocean depths permanently abyssal. We shall have occasion to return to this subject in the chronological geology.

SECTIONS ILLUSTRATING VARIOUS ARRANGEMENTS OF STRATIFIED ROCKS.

Fig. 38.—General Section in Southern New Brunswick.

A, Lower Carboniferous Conglomerate.

B, Mispeck Group,
C, Cordaite Shales,
D, Dadoxylon Sandstone
E, Conglomerate Group,
} Devonian.

F, St. John Group,—Cambrian.
G, Coldbrook Group,—Huronian?
H, Portland Group,— Laurentian?

Fig. 39.—Section at Walton, Nova Scotia. *a, a* Triassic Red Sandstone ; the lowest bed a hard calcareous conglomerate. *b.* to *f*, Lower Carboniferous. *b*, Red shale and marl with limestone bands. *c*, Gray and red conglomerate. *d*, Black shales. *e*, Small beds of limestone. *f*, Black shales with calcareous bands.

Fig. 40.—Ideal section representing the present arrangement of the Coal formation of Southern Pictou, Nova Scotia. *a*, Devonian ridge ; *b*, conglomerate ; *c*, great Coal measures ; *d*, minor Coal measures ; *e*, Lower Carboniferous ; *f*, Silurian.

(16.) GEOLOGICAL MAPS AND SECTIONS.

The facts and generalizations obtained on the above grounds are represented to the eye on maps and sections. On the former are indicated by spots or lines of colour or by differences of shading the different formations and their precise boundaries as far as ascertained. To these may be added marks indicating dip and strike and other arrangements. (Fig. 41.)

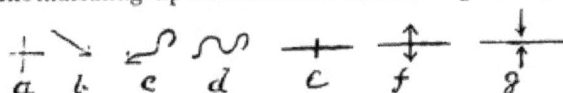

Fig. 41.—Marks used in mapping. *a*, Horizontal beds. *b*, Inclined. *c*, Undulating. *d*, Contorted. *e*, Vertical. *f*, Anticlinal. *g*, Synclinal.

While maps exhibit the horizontal distribution of formations, sections show more clearly the relations of age and superposition. Lines of section observed on the ground may in the first instance afford materials for the construction of a map, and when a map has been drawn these lines may be marked on it, and the sections along these lines may be drawn to accompany it. Such sections are usually named horizontal sections. But vertical sections may be obtained in shafts and borings, or may be constructed with the aid of the horizontal sections. (Figs. 38, 39, 40.)

The colours designating the different formations on Geological maps have usually been adapted to the geological features of particular countries. They should be so distinct from one another as to be readily distinguished either by natural or artificial light, and should as far as possible present some approach to harmony pleasing to the eye. One colour being selected for each system of formations, the sub-divisions may be indicated by shades of this tint, by white bars or dots traversing the tint, or by letters printed on it, or by combinations of these methods.

In the maps of the Canadian survey the following tints have been employed :—

Igneous Rocks	Shades of deep Red.
Laurentian	Carmine.
Huronian	Reddish Brown.
Cambrian, Siluro-Cambrian, Silurian	Shades of Slate.
Erian, or Devonian	Brown.
Carboniferous and *Permian*	Grey.
Triassic	Light Red.
Jurassic	Blue.
Cretaceous	Shades of Light Green.
Tertiary	Shades of Yellow.

In detailed maps Limestones are deep blue and Coal-beds black.

The colours recently proposed by the International Congress of Geologists, for the Geological map of Europe, are similar to the above, except that the Trias is violet and the Cambrian and Silurian greenish gray, both of which are objectionable colours in artificial light.

(17.) Field Geology.

Geological observation may be carried on in several different ways, the most important of which may be stated under the following heads :—

(1.) The Local Geologist.—It is scarcely too much to say that absolute ignorance of the structure and history of the earth, and more especially of the geology of the district in which we reside, is scarcely compatible with the mental health of any educated man, as it is certainly quite inconsistent with any intelligent comprehension of the geography and resources of our country. It is also true that the study of Geology from its wide range in time and space, and its intimate relations with physical geography, chemistry, physics, and the natural sciences generally, is well adapted to strengthen and cultivate our powers of observation, comparison and generalization.

Independently, however, of these subjective aspects of the science, a local amateur geologist often possesses facilities to make original discoveries and to extend the boundaries of knowledge. In order to do this, independently of some preliminary acquaintance with the principles and methods of the science, he should acquaint himself with what has been previously done in the geology of his locality, and should provide himself with a good topographical map, hammers, a compass and clinometer, a pocket lens ; and if possible with a microscope, a blow-pipe and its accessories, and a few of the ordinary chemical reagents; and he should keep a note-book wherein to enter all the facts observed. A cabinet or chest of drawers or of trays for specimens will also be desirable. So provided, he can, as opportunity offers, examine, draw and note down the various rock exposures in his vicinity, can collect specimens of minerals, rocks and fossils, and can watch for such temporary facilities for obtaining information as may be afforded by excavations, land-slips or accidental discoveries. In this way much may be done that could not be effected by a temporary survey, however detailed, and many precious facts and specimens may be preserved that otherwise would be lost. The local collector will soon become known to quarrymen and others, and specimens will be brought to him that otherwise would be neglected. It will be useful for him to establish relations with specialists who may aid him in the determination and description of new specimens, and such persons will usually be glad to do so if supplied with duplicates of the specimens. Useful exchanges may also be made with such persons. The local collector should beware of giving away unique specimens to mere tourists or curiosity hunters who will make no good use of them. Duplicates only should be given in this way, and usually such specimens alone as are not likely to be of importance afterwards. Many of the most interesting and important discoveries in the science are due to the labours of local geologists.

(2.) The Tourist.—Every traveller who wishes to understand the topography, scenery, productions, history and modes of life of the countries which he may visit, should know something of geology, and if he has opportunities to travel in districts not previously explored, he may be able to obtain most important facts respecting their structure. In order to do this he should be provided with the simple apparatus already mentioned, and above all with a geological map, should such exist, of the district and regions adjacent, and should read what has been already written as to their geology. If he can so arrange his time and route, he should traverse the country in directions at right angles to the prevailing structure, and should visit all good exposures within reach of his line of travel, making collections and taking full and accurate notes. In this way, by the intelligent observation of geological travellers, much of the information we possess of large portions of the earth has been acquired, while on the other hand costly expeditions have been rendered of little practical or scientific value by the absence of geological observation.

(3.) The Geological Explorer—who may be either an amateur or an officer of a Geological Survey. He may be supposed to devote his whole available time to geological observation and collection. He must be provided with a good topographical map, or, if this do not exist, with surveying instruments that he may make measurements for himself. It will be his duty to visit every exposure, to make careful notes and measurements of all the phenomena, and to collect specimens. He must not only cross the country by the best lines of section, but must trace the boundaries of every formation, and map them as accurately as possible, and must study the structure and arrangement of the formations, their faults, disturbances, mineral veins, intrusive masses and dykes, unconformities and denudation. He should not be content to be a mere mechanical explorer, but should endeavour to understand the structure as he proceeds, in order that he may at least hypothetically supply defects in exposures, and may have his attention devoted to questions which he might otherwise overlook. Work of this kind may of course admit of many degrees of detail, according to the facilities of travel and observation, and the time that can be devoted to it. It is the business of the geological explorer rather to work out the structure of the country than to search for useful minerals or fossils ; but he should give special attention to these whenever opportunity occurs, and should be particularly careful to mark and note every specimen in such a manner that no mistake may be made as to its true position, geological or topographical. Lastly, the geological observer should be prepared to note all facts connected with or dependent on the structure of the country, which may be useful economically or in relation to allied sciences.

(4.) The Mining Surveyor and Engineer —It is the object of the Mining Surveyor to discover and open up valuable mineral deposits ; but in order to effect this with advantage it may be necessary to do some preliminary work in geological exploration, in case this has not been previously done by others. In any case he should make himself personally familiar with the structure of the district, and with all the difficulties and complexities it may present, and with all good exposures which show the actual relations of useful minerals to the country rocks. He may have to make minute topographical surveys in connection with the courses of beds or veins, and may also have to conduct exploratory works, as trial-holes, trenches, adits or shafts, and to collect such samples as may be sufficient for assays and practical trials of their value.

The Civil Engineer, concerned in preliminary surveys for engineering works, or in actual construction, is also interested in the geological structure of the country in which he works. The distribution of geological formations regulates the contours and the nature and position of the rock masses under the surface. Hence the cost of excavation and construction, the stability of structures and the materials of which they may be composed and the accidents to which they are liable, are directly dependent on the geological formation of the country ; and difficulties may be avoided and advantages secured by the judicious application of even a small amount of geological knowledge. In like manner, the miscalculations, errors in specification and unexpected difficulties and pecuniary expenditures arising from want of attention to geological facts, are fruitful causes not only of loss but of dispute and litigation. Such difficulties may arise either from inattention to the facts, or from ignorant attempts to specify details not properly understood. Other things being equal, he will be the best and most economical engineer who best understands the rocks in or on which his operations are being conducted.

PART II. HISTORICAL GEOLOGY.

CLASSIFICATION AND TERMS.

The application of the facts and principles of lithology, stratigraphy and palæontology to any given district, enables us to work out the geological succession of formations or geological history of the district in question, including not only the physical changes but the changes in living beings that may have occurred. The comparison and grouping of such local results enables us at length to frame a table or chart of the geology of the whole earth. This we shall now proceed to construct, beginning with the oldest formations, and giving, wherever practicable, typical examples of each from Canada, or from those regions in America or elsewhere in which it may be best developed and has been most fully studied.

The whole geological history of the earth may be included in four great Periods or Eras, the names of which have been based on the progress of animal life. They are, beginning with the oldest—

1. *The Eozoic,** or that of Protozoa, often called Archæan.

2. *The Palæozoic,* or that of Invertebrate animals.

3. *The Mesozoic,* or that of Reptiles.

4. *The Kainozoic,* or that of Mammals and of Man.

They are farther subdivided into *Ages,* or if we regard the rocks themselves rather than the time occupied in their deposition, into *Systems of*

---

* "Azoic," the term originally proposed by Murchison is not now applicable, and the same objection applies to the term "Agnotozoic" recently proposed by western Geologists. The only objection to "Archæan" is that it does not carry out the idea of succession of life embodied in the names of the other periods.

Formations. These are represented in the following table, beginning as before with the oldest :—

Eozoic or Archæan. { Laurentian.
Huronian.

Palæozoic........ { Cambrian.
Siluro-Cambrian.
Silurian.
Erian or Devonian.
Carboniferous.
Permian.

Mesozoic. .... { Triassic.
Jurassic.
Cretaceous.

Kainozoic........ { Eocene.
Miocene.
Pliocene.
Pleistocene.
Modern.

Systems may be divided into Series, and these into Stages or Subdivisions :—These again into beds. The method recommended by the International Congress is as follows :—

1. *Groups* or *Eras* ; *ex.* Palæozoic.
2. *Systems* or *Periods* ; *ex.* Silurian.
3. *Series* or *Epochs* ; *ex.* Niagara.
4. *Stages* or *Ages* ; *ex.* Niagara shale.

In the Reports of the Geological Survey of Canada—

Systems are divided into Groups,
Groups into Formations,
Formations into Series,
Series into Beds.

In noticing the Systems of formations and their subdivisions in detail, we shall begin in each case with a general statement of the subdivisions of the system and their most characteristic fossils, more especially noting the earliest known appearance of each leading animal and vegetable type. We shall then describe some typical Canadian locality, should there be such, and the distribution of the system in Canada and elsewhere. Finally, we shall mention characteristic useful or economic minerals.

Since Canada embraces about half the area of North America, and includes portions of all the geological formations of the continent, we shall in most cases be able to obtain within its limits typical examples of rocks and fossils; and when these fail, shall have recourse to other regions.

EOZOIC PERIOD.

(I.) LAURENTIAN SYSTEM.

1. *Lower Laurentian* or Ottawa Series. In Canada—Orthoclase gneiss of Trembling Mountain (Logan), Ottawa gneiss (Geol. Survey), lower part of Lower Laurentian, of Logan. European equivalents—Bogian gneiss, Ur gneiss, Lewisian gneiss. Consists mostly, so far as known, of beds of Orthoclase gneiss destitute of fossils, constituting the "fundamental gneiss" of some geologists. In this and the next group there is much porphyritic gneiss (the Augen Gneiss of Scandinavia.)

Fig. 42.—Section showing the mode of occurrence of *Eozoon* in Middle Laurentian a St. Pierre. *(a)* Gneiss; *(b)* Limestone; *(c)* Diorite and Gneiss.

2. *Middle Laurentian* or Grenville Series. In Canada—Gneiss, diorite, limestone, pyroxene rock, &c., of Grenville, Petite Nation, &c., being the upper part of the Lower Laurentian of Logan. European equivalents—Ur gneiss in part, Lewisian gneiss in part, Etage A of Bohemia in part, Dimetian of Wales? Gneiss and crystalline limestone of Brittany. In the upper part of this series there is much quartzite and garnetiferous gneiss, also ferruginous gneiss holding magnetic grains and beds of Magnetite.

EOZOIC FOSSILS.—*Eozoon Canadense*, (Figs. 43 to 44) also graphitised plants.

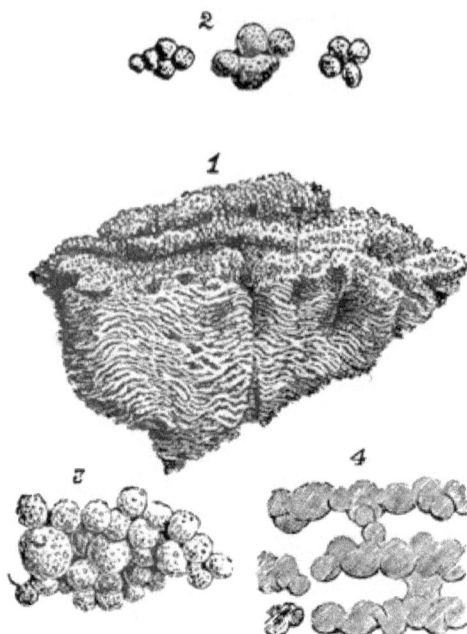

LAURENTIAN FOSSILS.

Fig. 43.—*Eozoon Canadense.*   (1) Small specimen disengaged by weathering. (2) Acervuline cells of upper part—magnified. (3) Tuberculated surface of laminæ—magnified. (4) Laminæ of Serpentine in section, representing casts of the sarcode—magnified.

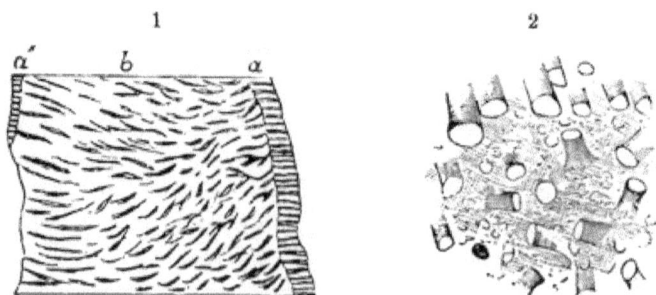

Fig. 44.—Structures of Eozoon. (1) Section magnified, showing tubuli at *(a)* and canals at *(b).* (2) Canals more highly magnified.

3. *Upper Laurentian,* (*Norian* of Hunt.)   In Canada—Labradorite and Anorthosite series of the Ottawa district, &c.   European equivalents —Etage A of Bohemia in part, Dimetian of Wales? Norite formation of Scandinavia.   No fossils known.   In the district originally described by Logan there occur large masses of Labradorite rocks, now believed to be intrusive ; but there and elsewhere there are gneisses, schists, &c., and the formation is usually characterised by a prevalence of basic and lime felspar as distinguished from the orthoclase of the Laurentian.   Hence bedded Anorthosite rocks and gneissic Anorthosite are characteristic features.   To this horizon may belong the Limestones, Schists and bedded Dioritic rocks of Southern New Brunswick, which underlie the Huronian in that district ; also the "White Mountain Series" of Hitchcock occupying a similar position in New Hampshire and the Green Mountains, and the similar rocks called Upper Laurentian by Kerr, in South Carolina. A felspathic and Mica Schist group of this age seems indeed to be very generally associated with the typical Laurentian of Eastern America.

The Laurentian group of rocks has been recognized in Brazil and elsewhere in South America.   In the Old World it occurs in Scandinavia, in the Western Highlands of Scotland, in Brittany, in Bohemia and elsewhere in Eastern Europe, and in Central and Eastern Asia.   It recurs in Arabia and in Africa, extending from the first cataract of the Nile far to the southward.   The Laurentian may indeed be regarded as a universal foundation of the Continents ; though in many very large areas buried under newer sediments.

In the Laurentian districts there are great masses and veins of Granite, binary Granite, Diorite, Labradorite rock and Dolerite.   Some of these seem to be either contemporaneous or little newer than the containing rock. Others have been introduced at much later periods.

Distribution in Canada, &c.   These formations constitute an extensive angular belt extending south-westward, north of the St. Lawrence valley, from Labrador to the Western coast of Lake Superior and thence north-west to the Arctic ocean, and they recur in Greenland.   At the Thousand Islands this belt is connected with an extensive peninsula in the State of New York.   Minor areas protrude through the Palæozoic rocks in New-foundland, New Brunswick, and the Atlantic coast of the United States, and also probably in the mountainous belt fringing the Pacific coast.

The following Section given by Logan on the North Side of the Ottawa may be regarded as characteristic.*

---

* Geology of Canada, p. 45.

Section from Trembling Mountain, in the County of Ottawa.
(Logan).—

(Order ascending.)

First Orthoclase Gneiss of Trembling Mountain,
    (Lower or Ottawa Gneiss).............   5,000 feet, or more.

First Limestone, or Limestone of Trembling Lake.   1,500

Second Orthoclase Gneiss, between Trembling
    Lake and Great Beaver Lake...........   4,000

Second Limestone, or Limestone of Great Beaver
    Lake and Green Lake, with two interstrati-
    fied bands of garnetiferous rock and horn-
    blendic orthoclase gneiss, making up about
    half its volume......................   2,500

Third Orthoclase Gneiss, with bands of garnet-
    iferous gneiss and Quartzite, between Beaver
    Lake and the Rouge River.............   3,500

Third Limestone, or Limestone of Grenville, in
    some places including a band of Gneiss:
    (Eozoon Canadense). Its thickness varies
    from 1,500 to 60 feet, average thickness
    estimated at........................   750

Fourth Orthoclase Gneiss, including a thin bed of
    Limestone (Proctor's Lake), and 600 feet of
    Quartzite..........................   5,000

Norian or Labradorian, or Upper Laurentian Series,
    estimated at........................   10,000
    _____
                    Total...............32,250

The genesis of the Laurentian rocks has been a subject of much discussion. To the writer it has long appeared that the lowest or Ottawa

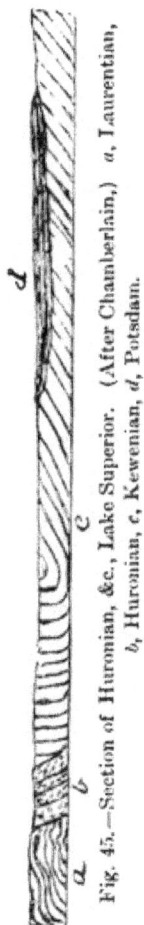

Fig. 45.—Section of Huronian, &c., Lake Superior. (After Chamberlain.) a, Laurentian, b, Huronian, c, Kewenian, d, Potsdam.

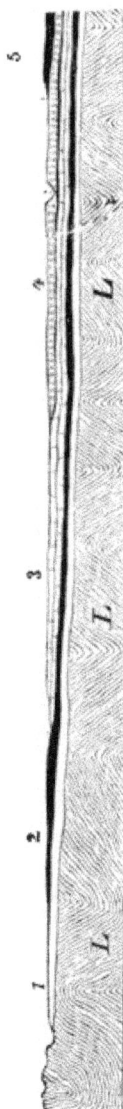

Fig. 46.—Superposition of Cambrian and Silurian on Laurentian (L) in Quebec and Ontario. (1) Cambrian. (2) Siluro-cambrian. (3) Silurian. (4) Devonian. (5) Carboniferous.

Gneiss is an aqueo-igneous product of the original ocean covering the cooling crust. The Middle Laurentian, while presenting the same characters in part, introduces also the formation of sandstones, mud-rocks and limestones, with organic deposits. The Upper Laurentian marks a transition period, with varied local deposits, igneous and aqueous, indicating the beginning of those great movements of the crust which closed the Laurentian era.

Economic Products.—In Canada the Laurentian abounds in Gneiss, Syenite, Crystalline Limestones, Serpentine and other rocks suitable for construction and for ornamental purposes. It contains beds of Magnetite (Hull, Marmora, &c.) and veins of Hematite (Perth), and the ferruginous gneisses have yielded Iron-sand, (Moisic, &c.). Important veins of Apatite occur, especially in the Pyroxenic beds and the Micaceous schists, (Buckingham, Templeton, Burgess, &c.), Graphite occurs in large quantity (Buckingham, &c.), Mica, Galena, Gold, (Madoc, &c.) are also locally among the products of the Laurentian.

(II.) HURONIAN SYSTEM.

1. *Huronian proper.*—In Canada—Chloritic slate, jasper conglomerate, slate conglomerate, quartzite, limestone and bedded diorite of Georgian Bay. Similar rocks in Newfoundland, New Brunswick and possibly also in the Eastern Townships of Quebec. European equivalents—Urschiefer of Scandinavia, Etage A of Bohemia and Pebidian of Wales (Hicks).

FOSSIL.—*Eozoon Bavaricum*, Gumbel,—a somewhat doubtful form— (Fig. 46). The Huronian Limestones have hitherto afforded no fossils in Canada, though I have detected traces of spicules, probably of sponges in the chert contained in them.

The following section (Logan, *Geology of Canada*) represents the structure of the Huronian in the typical region of Georgian Bay, in ascending order :—

1. Green silicious and slaty strata with bluish and black slates.
2. Slate conglomerates, or slaty or earthy rock, probably in part volcanic ash, with numerous pebbles of Laurentian rocks; in and near the base greenish slates.
3. Impure Limestone, with chert. (The chert shows traces of silicious spicules.)
4. Slate conglomerate, greenish slate and Quartzite.
5. Quartzite, with quartz conglomerate and jasper conglomerate.

Associated with these beds are extensive sheets of stratified Diorite ; and in places they are traversed with Diorite and Syenite Dykes. The Copper veins of the Huronian district of Georgian Bay attain their maximum thickness in the bedded Diorite.

The thickness of the Huronian on Georgian Bay is estimated at 18,000 feet, but this is probably only a part of the volume of the entire system.

The " Hastings Group," of Ontario, may be of Huronian age, but it is
not improbable that some of the crystalline schistose rocks west of Lake
Superior usually classed as Huronian may really be upper Laurentian.

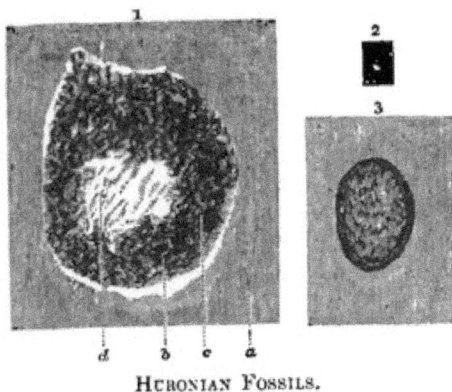

HURONIAN FOSSILS.

Fig. 47.—(1.) Cast of worm burrows, Madoc (Hastings group)—magnified. *a*, Containing rock. *b*, Space filled with calcite. *c*, Sand agglutinated and
stained black. *d*, Sand uncoloured. (2, 3,) Another specimen, natural
size, and magnified.

Fig. 48.—*Eozoon Bararicum* × 25 (after Gumbel). *a*, *b*, calcite. *c*, tubuli.
d, *e*, Casts of contorted chambers.

Dr. Selwyn describes the Laurentian and Huronian succession on the
North Shore of Lake Superior, on the line of the Canada Pacific Railway, as follows, in ascending order :—

Laurentian.

1. Red, gray and white Orthoclase Gneiss in great variety.
2. Black Hornblendic Gneiss, often Garnetiferous, and cut by veins of
red and white Orthoclase.
3. Pyroxenic Gneiss, laminated.
4. Crystalline Limestone in thick beds.
These beds are all much contorted but distinctly stratified.

Huronian.

1. Felsites and Felsitic Quartzites, red and white.

2. Dark coloured Quartzites and dark gray and black silicious beds, the lower part holding angular fragments and pebbles of white granite or gneiss.

3. Diorite and Diabase, with chloritic or hornblendic ash-rocks and agglomerates.   Associated with these are bands of argillite.

These Huronian rocks are overlaid by the Vermillion River Grey Sandstones or Quartzites and Slates, supposed to be of Cambrian age.[*]

2. *Upper Huronian* or *Lowest Cambrian.*   In Canada—Conglomerates, slates and grits of Eastern Newfoundland, Intermediate or Signal Hill beds; Animiké and Kewenian group of Lake Superior.   Upper Huronian of Southern and E. New Brunswick.   European equivalents not certainly known.   This must be regarded for the present as a provisional group. It may include the "Intermediate Series" of Murray, in Eastern Newfoundland, the upper part of the Pre-Cambrian rocks of Southern New Brunswick, the Animiké and Kewenian rocks of the basin of Lake Superior, and possibly some of the rocks called Taconian by Hunt.   These rocks may, on the other hand, be ultimately referred in part to the Lower Cambrian and in part to the true Huronian.   The following section shows the development of the Kewenian at Maimanse, Lake Superior,[†] order ascending :—

1. Conglomerate with granitic and gneissic pebbles and boulders—the latter sometimes two feet in diameter,—about 250 feet.

2. Crystalline and amygdaloidal trap, (dolerite) with conglomerate; veins of quartz and calcite with native copper and copper ores,—about 700 feet.

3. Argillo-arenaceous beds and mottled argillaceous sandstone,—about 350 feet.

4. Alternations of trap and tufa, with a bed of conglomerate,—veins of calc-spar, quartz and laumonite, with native copper and silver,—about 750 feet.

FOSSILS.—*Aspidella Terra-novica*, Billings ; *Arenicolites.*

Fig. 49.—*Arenicolites Spiralis.*   Fig. 48.—*Aspidella terranovica*—Billings.
Upper Huronian of Newfoundland.

[*] "Descriptive Sketch," 1884.
[†] Paper by the Author, Canadian Naturalist, 1857.

Distribution.—The Huronian is extensively developed on the north side of Lake Huron and south and west of Lake Superior. It occurs also in Newfoundland and New Brunswick, and probably in various parts of eastern Quebec and the Atlantic States.*

The Huronian in the typical region of Georgian Bay and also on Lake Superior, and in New Brunswick and Newfoundland is mainly a littoral deposit, formed along the margin of the ancient Laurentian land, and indicating extensive movement of stones, boulders, sand and mud, not improbably in part by ice action, along with great igneous ejections of Diorite, &c., consequent on the foldings of the crust which closed the Laurentian age. The deep-water formations of the period are little known to us ; but their margins may be represented by the limestones and black slates. The forms of Protozoa resembling Eozoon, the burrows of worms and remains of sponges indicate oceanic conditions.

Useful Minerals.—The Huronian includes the great copper veins of Georgian Bay and the gold and silver deposits of the western extremity of Lake Superior, the iron ores of Marquette and Menominee. The Trappean ash rocks and gabbros of Mainanse, Michipicoton Island and Kewenaw Point belong to the Kewenian, and contain much native copper and silver. The rich silver deposits of Silver Islet, Lake Superior, belong to the Animiké formation.

PALÆOZOIC PERIOD.

(I.) CAMBRIAN SYSTEM.

1. *Lower Cambrian.*—In Canada the quartzite and slate of the Atlantic coast of Nova Scotia (the gold series,) containing *Astropolithon* and trails or markings known as *Eophyton* and *Scolithus*, may be referred to this age. We may also place here the Basal series of Matthew in Southern New Brunswick, consisting of conglomerate, sandstone and shale with linguloid shells, *Obolus pulcher*, Matthew, and worm burrows, *(Psammichnites* and *Scolithus.)* To the upper part of it are now referred the Georgia beds of Vermont and Quebec with *Olenellus Thompsoni.* Lapworth has (1888) announced the discovery of *Olenellus* and its associates in the Lower Cambrian of England. It would appear that in Europe and America as many as 67 genera and 165 species of marine invertebrates, including all the leading types of marine life, have been found in the Lower Cambrian. †

---

* The precise relations of the Hastings group of Eastern Ontario, and various other groups of altered rocks resembling it in mineral character, with the Huronian, are not yet well understood.

† Walcott.

E

In America, in addition to *Olenellus*, *Ptychoparia*, *Microdiscus*, *Ellipso-cephalus*, *Hyolithes*, *Obolus*, *Kutorgina*, *Eocystites*, *Leptomitus* and other genera have been found.

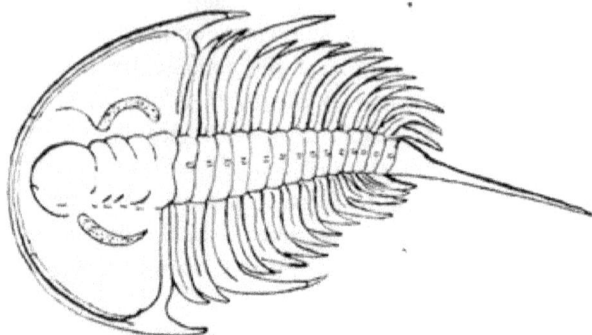

Fig. 50.—*Olenellus Thompsoni*, Georgia slates. (After Walcott.)

Here some would also place provisionally, in the absence of fossils, the Kewenian, or Upper Copper bearing series of Lake Superior, and the Signal Hill series of Newfoundland.   Both of these are formations of conglomerate and sandstone of reddish colours and associated with igneous rocks and resting on the Huronian.   It is possible, however, that these beds may constitute another series below the Cambrian.   They may correspond with some of the older sandstones and schists of Western Europe which underlie the fossiliferous Cambrian.

The Caerfai series of Hicks in Wales, consisting of purple, red and green sandstones, shales and conglomerates, seem to be the equivalent of these beds, and also etages A and B of Bohemia, and the Eophyton and Fucoidal shales of Sweden.   The known rocks of this age are everywhere littoral or coast formations.

2. *Middle Cambrian.*—In this we may place the Acadian series of New Brunswick, Division 1 of Matthew in the St. John group, consisting of slates with abundant fossil remains, and equivalent to the Solva and Menevian groups of Wales, and to the Paradoxides beds of Sweden. They are also the equivalents of the Paradoxides slates of Southern Newfoundland and of Braintree in Massachusetts.   These beds are divided by Matthew into four series, (a to d) of which c and d are those to which the name Acadian group was originally given.   Above these at St. John are other slates containing fossils which indicate a somewhat higher horizon passing into the Upper Cambrian.

FOSSILS.—*Paradoxides*, (Fig. 51.) *Conocoryphe, Ctenocephalus*, (Fig. 52.) *Microdiscus, Agnostus,* and other trilobites, also *Lingulella*, (Fig. 53) *Orthis*, (Fig. 54.) *Stenotheca*, (Fig. 55.) *Hyolithes, Eocystites, Protospongia, &c.* In the upper Division are *Ctenopyge, Ptychoparia, Agnostus, Kutorgina, &c.* The species and many of the genera are distinct from those in the lower group.

Matthews has catalogued 65 species from this series, of which 28 are Trilobites, but the Mollusks, Echinoderms, Coelenterates and Protozoa are also represented.

At Mount Stephen in the Rocky Mountains, on the line of the Canada Pacific Railroad, fossils of Lower and Middle Cambrian age have been found in a formation of quartzite, slate and limestone.

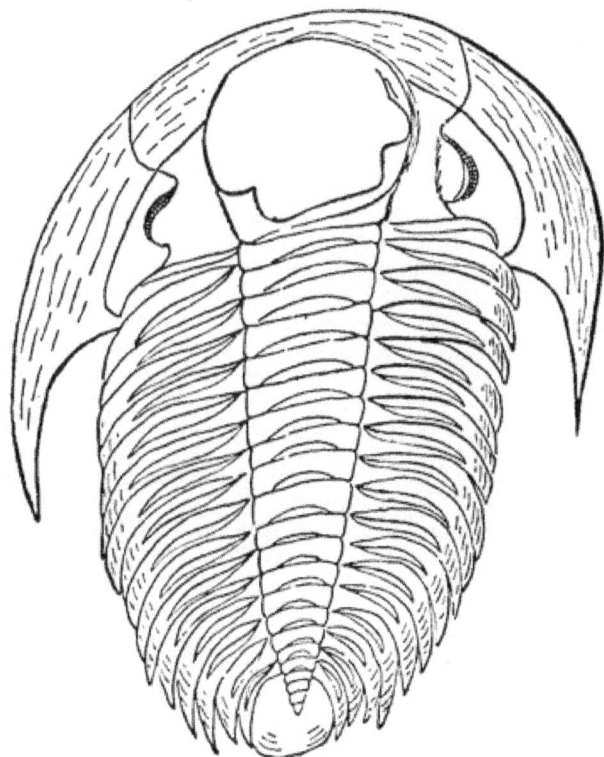

Fig. 51.—One-fourth natural size.

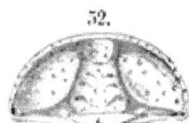

MIDDLE CAMBRIAN FOSSILS.

Fig. 51.—*Paradoxides Regina.* 52. *Ctenocephalus Matthewi.* 53. *Lingulella Matthewi.* 54. *Orthis Billingsi.* 55. *Stenotheca Acadica.*—Acadian group, St. John, N.B.

3. *Upper Cambrian.*—The characteristic formation of this age in Canada is the Potsdam sandstone, to which may be added the lower part of the Calciferous formation. It is widely spread over the lower St. Lawrence and in the State of New York, and is remarkable for the abundance of the cylindrical burrows known as *Scolithus,* for tracks of crustaceans (*Protichnites, Climacichnites,*) and for species of *Lingulella* and Trilobites of the genera *Olenus* and *Dikellocephalus.* It is the equivalent of the Lingula Flags of Wales and the Olenus Zone and Dictyonema Slates of Sweden, the latter being equivalent to the English Tremadoc and the Lower Calciferous of Canada.

The Miré River Slates of Cape Breton, Flags of Kelly Island, Newfoundland, Upper Slates of St. John, New Brunswick, Limestones of L'Anse-à-Loup, Labrador, and the *Dictyonema* shales of Matane and C. Rosier belong to this or in part to the previous group, and various Corals, Crinoids, Lamellibranchs, Heteropods, Gasteropods and Cephalopods occur in the upper member, which shows transition to the next age.

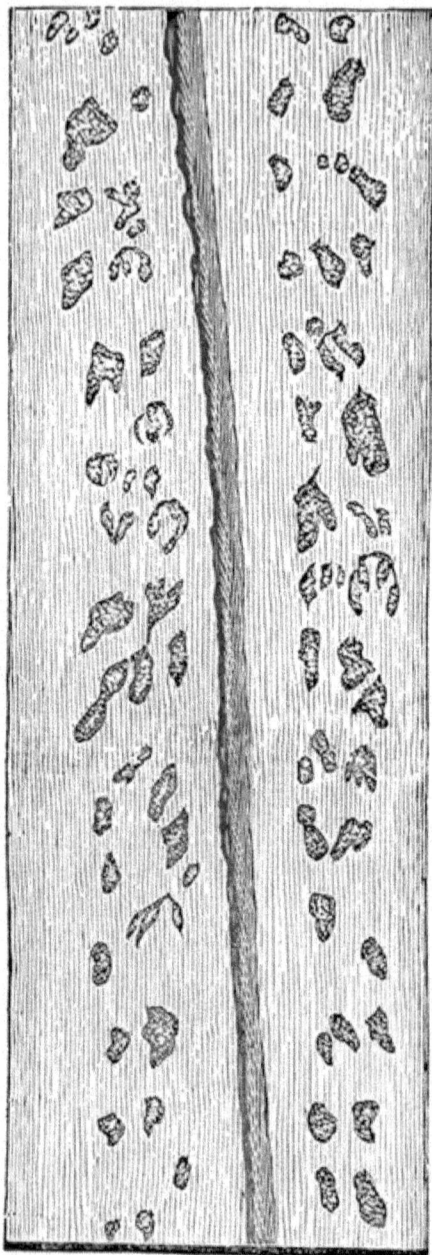

Fig. 56.—*Protichnites septem-notatus*, Owen, Potsdam.

This Upper Member, the Calciferous sand-rock of the original survey of New York, is represented by coarse dark-colored dolomite over the continental plateau, but contains slaty beds *(Dictyonema* shales*)* on the Lower St. Lawrence. It may admit of division into two members belonging respectively to the Cambrian and Siluro-Cambrian.

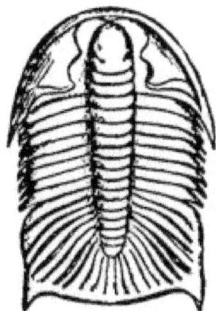

Fig. 57.—*Dikellocephalus Minnesotensis.*

Fig. 58.—*Lingulella antiqua* *c*, Short variety. *d*, Long variety.

Distribution.—The Lower Cambrian is best developed on the Atlantic coast of Nova Scotia, where it constitutes the gold-bearing series. It occurs also in southern New Brunswick and in New England and New York as well as in the extreme west. The Middle Cambrian, or Paradoxides series seems in great part limited to the Atlantic coast, not having been found in the interior continental plateau of America. The most typical Canadian representatives are those of the vicinity of St. John, New Brunswick. It occurs, however, also in the west. The Upper Cambrian is found in Quebec and Ontario, and in the State of New York, and also in the Rocky Mountains; but appears to be less developed on the Atlantic margin. The Upper Cambrian is extensively represented by the Potsdam Sandstone and Calciferous sand-rock all over the continental plateau of America. It would thus appear that the Middle Cambrian begins with a period of continental elevation followed by gradual subsidence. Similar evidence exists in Europe. The Potsdam is extensively developed in the western part of the Province of Quebec, and in New York, where it often rests directly on the Laurentian.

Minerals.—The gold veins of Nova Scotia occur in Cambrian rocks, especially at and near the junction of the quartzite and slate constituting the Lower Cambrian of the Atlantic coast. They are best developed on the anticlinals, and often coincide in direction with the strike of the beds. The matrix is usually quartz, in which the gold occurs either disseminated in microscopic grains or in nuggets. These veins are of later age than the beds they traverse, probably as late as the Erian or Devonian. In some parts of the coast region of Nova Scotia Staurolite, Chiastolite and Garnet have been developed in the altered slates.

In the Eastern Townships of the Province of Quebec, where large areas at one time supposed to be Siluro-Cambrian (Quebec group) are now mapped as Cambrian and Pre-Cambrian by the Geological Survey, there are extensive deposits of Copper, (Harvey Hill, Wickham, &c.,) also veins of Gold, Chromic Iron, Manganese, Fibrous Chrysotile (Asbestos), and Soapstone. These deposits are believed to occur principally in the Cambrian and Pre-Cambrian areas.

(2.) SILURO-CAMBRIAN SYSTEM.

(Lower Silurian of Murchison, Ordovician of Lapworth.)

1. *Quebec Series* (Calciferous-chazy).—In Canada &c., shales, limestones and sandstones of Point Levis, and the south side of the Lower St. Lawrence, Upper Calciferous Dolomite and Chazy Limestone of the interior plateau. In the United States a belt extending southward through Vermont and New Hampshire. European equivalents—Llandeilo series in part, Arenig (Skiddaw and Borrowdale) of England; Etage D1, Bohemia; Lower Graptolithic slates of Scandinavia.

The Quebec Group proper is a submarginal series belonging to the Atlantic border, and from its coastal position has been subjected to great lateral pressure and folding, and locally to metamorphism. The equivalent rocks of the continental plateau are more calcareous and less disturbed, and their fossils are for the most part different. The disturbed position and peculiar conditions of deposit of the Quebec group have caused much perplexity, and this is increased by the ancient facies of the fossils as compared with those of the plateau deposits. It is also now known that rocks and fossils extending from the Upper Cambrian to the Trenton group have been folded in with the beds of the Quebec group proper.

FOSSILS.—Graptolites of Genera *Graptolithus, Phyllograptus, Dendrograptus, Dichograptus, Dictyonema,* &c. Trilobites of Genera *Dikellocephalus, Agnostus, Arionellus, Bathyurus,* &c., Land plants—*Protannularia* of Skiddaw series.

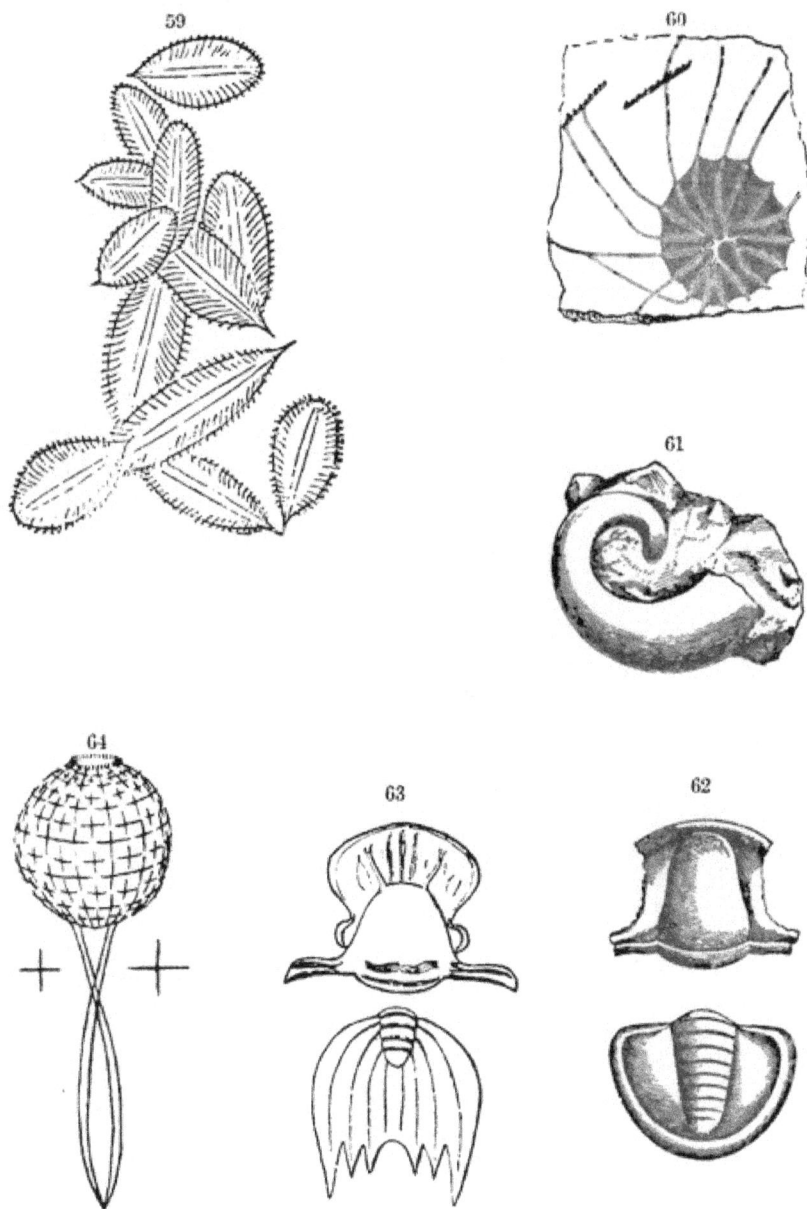

SILURO-CAMBRIAN FOSSILS.   (Quebec group.)

Fig. 59.—*Phyllograptus typus.*   60.—*Dichograptus Logani.*   61.—*Ecculiomphalus intortus.*   62.—*Bathyurus Saffordi.*   63.—*Dikellocephalus magnificus.*   64.— *Protospongia tetranema.*

The Graptolites of the Quebec group have been described by Prof. James Hall, and more recently comparisons have been instituted between the succession of genera and species in Canada and in Europe by Prof. Lapworth, from which it appears that the oldest graptolithic beds (Matane, Cape Rosier, &c.) containing *Dictyonema sociale*, are, as stated by the writer in 1883, as old as Tremadoc, the Levis graptolites, (*Phyllograptus*,) *Tetrograptus*, *Dichograptus*, &c. represent the English Arenig, and certain upper beds, (Marsouin, &c.) are of Trenton or Utica age. This graptolithic succession may be stated thus, in ascending order :—

1. *Dictyonema Zone*, holding *D. sociale*, Salter, equivalent to Tremadoc and Calciferous.

2. *Phyllograptus Zone*, holding *P. typus*, &c., and Levis trilobites and *Protospongia*, equivalent to Arenig and Upper Calciferous or Chazy. Typical Quebec group of Logan.

3. *Cœnograptus Zone*, holding *C. gracilis* and Trenton fossils. Equivalent to English Bala and to Black River and Trenton.

All of the above are found at various points in the long range of Cambrian and Ordovician deposits between Cape Rosier and Quebec and thence south-westward, and which collectively constituted the Quebec group of Logan.

4. *Diplograptus Zone*, with *D. pristis*, &c., equivalent to Upper Caradoc and Utica shale. Beds of this age occur on the North shore of the St. Lawrence near Murray Bay and Les Eboulements, also at Lake St. John. Certain beds associated with the Quebec group at Cove fields and Orleans Island near Quebec show a transition toward this fauna, according to Lapworth.

The Graptolithic fauna is of special value because it is oceanic and not limited to continental plateaus. It presents identical characters in the west of America as well as in Europe, and even in Australia. An interesting illustration is furnished by the discovery of graptolites on the Dease river, in the Northern part of British Columbia, by Dr. G. M. Dawson. According to Prof. Lapworth they are mostly known species of Middle Ordovician age. Similar species were found by Mr. R. S. McConnell in the Kicking Horse Pass, Rocky Mountains.*

---

* Canadian Record of Science, 1888,

2. *Trenton Series.* In Canada—Black River and Trenton limestones of Quebec and Ontario. Corresponding rocks of the New York series. European equivalents—Bala formation of England and Wales; Etage D 2 of Bohemia; Regio C, or Oeland limestone of Scandinavia; Graptolite and Calymene slates of France—Second Fauna of Barrande.

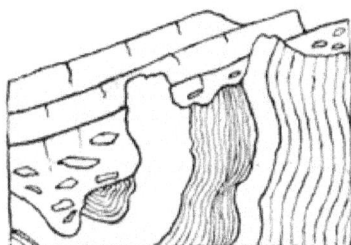

Fig. 64a.—Superposition of Siluro-Cambrian limestone on quartzite and slate of Hastings series, Hog Lake, Ontario.

The Trenton series represents a submergence of the whole continental plateau of North America under warm waters richly tenanted by a great variety of forms of invertebrate life, and representing the culmination of the invertebrate animals in the Lower Palæozoic. More especially the great Cephalopods of the family Orthoceratidæ were dominant at the time, and the genera *Asaphus, Calymene* and *Trinucleus* replace the older forms of Trilobites. The Utica shale overlying the Trenton represents an influx of cold and muddy waters into the great inland sea of the Trenton, along with the entrance of Graptolites and other oceanic forms of life.

FOSSILS.—Rich invertebrate Fauna of Corals, Crinoids, Brachiopods, Lamellibranchiates, Gasteropods and Crustaceans. The following are very characteristic in Canada—*Monticulipora* (different species), *Columnaria alveolata, Tetradium fibratum, Ptilodictya acuta, Glyptocrinus ramulosus, Strophomena alternata, Leptaena sericea, Orthis lynx, Lingula quadrata, Cyrtodonta Huronensis, Murchisonia bellicincta, Pleurotomaria subconica, Conularia trentonensis, Asaphus megistos, Trinucleus concentricus.*

In Nova Scotia and New Brunswick the Trenton and Quebec series appear to be represented by the Graptolite slates of Northern New Brunswick, and by the felsites, agglomerates, slates, &c. of the Cobequid

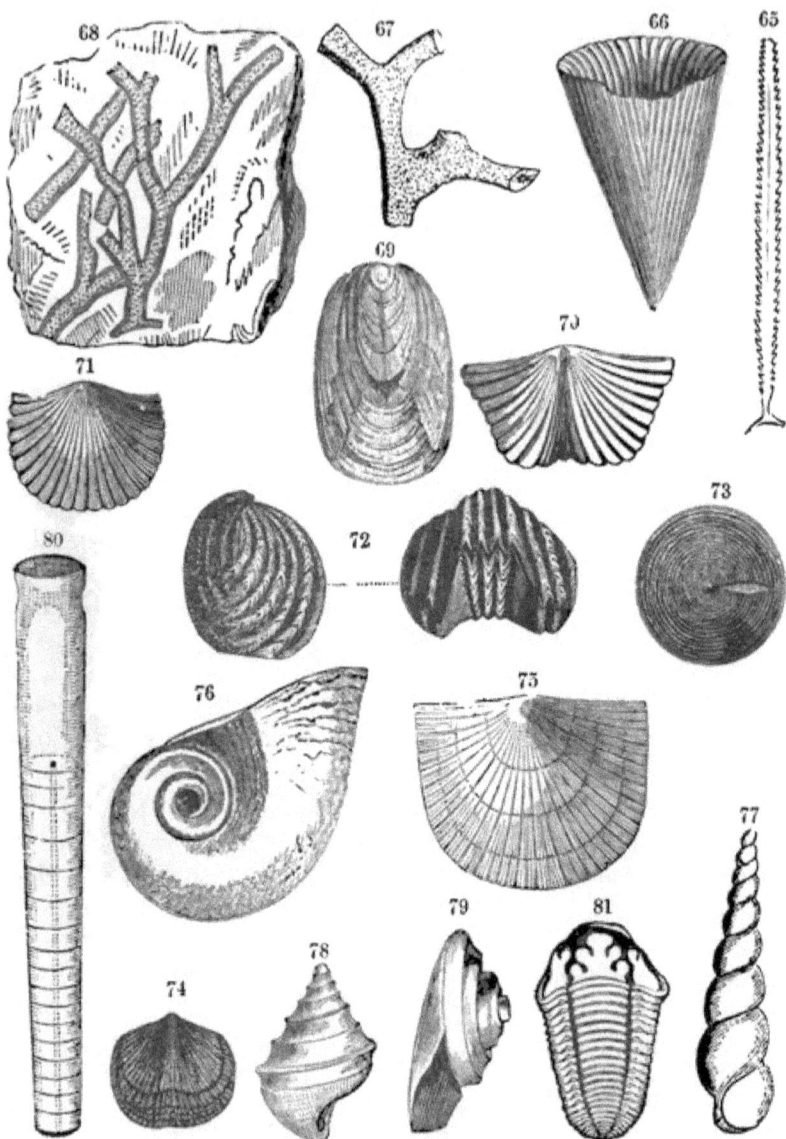

SILURO-CAMBRIAN FOSSILS.

Fig. 65.—*Graptolithus bicornis.* 66.—*Petraia profunda.* 67.—*Monticulipora,* sp. 68.—
Ptilodictya acuta. 69.—*Lingula quadrata.* 70.—*Orthis lynx.* 71.—*Orthis
pectinella.* 72.—*Rhynchonella increbrescens.* 73.—*Discina circe.* 74.—*Orthis
testudinaria.* 75.—*Stophomena alternata.* 76.—*Bellerophon Sulcatinus.* 77.
—*Murchisonia gracilis.* 78.—*M. bicincta.* 79.—*Pleurotomaria umbilicatula.*
80.—*Orthoceras* sp. 81.—*Calymene senaria.*

Mountains, &c. in Nova Scotia, which have been named the Cobequid series. They resemble in mineral character the Borrowdale series of England.

3. *Hudson River Series.* In Canada—Utica shale of the St. Lawrence valley, shales, coarse limestones and sandstones overlying the Utica in various parts of Ontario and Quebec, and extending southward into the United States. European equivalents—Caradoc sandstones and shale. Regio D of Scandinavia ; Etages D3, D4, Bohemia.

FOSSILS.—Continuation of invertebrate Fauna of Trenton in part, with some new types, as *Favistella stellata, Halysites gracilis, Pterinea demissa, Asaphus Canadensis, Triarthrus Beckii,* and *T. spinosus. Graptolites* abound in some parts of the Utica, especially *G. pristis, G. bicornis* and *G. ramosus.* Earliest certainly known land plants—*Protostigma, Annularia, Sphenophyllum.*

Distribution.—Formations of this age occur in patches along the northern coast of the Gulf and River St. Lawrence, on the north side of Anticosti and on the south side of the St. Lawrence from Gaspe. From Bay St. Paul upward, they occupy both sides of the St. Lawrence and the valley of the Lower Ottawa, as far up as the Thousand Islands. Westward of this they form a broad belt extending across Ontario, from Lake Ontario to Georgian Bay. They occur in the Islands of Georgian Bay and the North Channel, and at River St. Mary cross over into the United States. They reappear in the Valley of the Red River.

Useful Products.—The limestones of the Chazy and Trenton groups afford good building stone and lime. Sandstones and flags are found in the Hudson River series. The Utica shale is in some places sufficiently rich in bituminous matter to afford illuminating oil, and it is also a source of natural gas. The copper and other metallic minerals of the Eastern Townships of Quebec, formerly referred to the Quebec group, are now regarded as for the most part occurring in older rocks.

III. SILURIAN SYSTEM.

(Upper Silurian of Murchison.)

1. *Medina Series.* In Canada—Sandstones of the West end of Lake Ontario and extending thence into the United States. Lower part of Anticosti series. European equivalents—Llandovery formation of Wales or beds of passage, including the Mayhill sandstone. Etage E 1 Bohemia.

FOSSILS.—The trails known as *Arthrophycus,* and *Lingula Cuneata* are characteristic.

2. *Niagara Series.*   In Canada—Clinton and Niagara limestone of Ontario, and their extension southward into the United States.   Lower Arisaig and New Canaan slates of Nova Scotia ; Upper Silurian limestones and slates of Northern New Brunswick and Gaspé in part. European equivalents—Wenlock limestone and shale of England.   Etage E 2 of Bohemia.

FOSSILS.—The Niagara limestone contains a rich marine fauna; *Astylospongia praemorsa, Stromatopora concentrica,* and Corals, &c. of the genera *Favosites, Halysites, Heliolites, Dictyonema* ; Crinoids, as *Stephanocrinus* and *Caryocrinus* ; Mollusks, as *Strophomena rugosa, Pentamerus, Spirifer Niagarensis.*   Trilobites of genera *Dalmania, Lichas, Calymene,* and *Illaenus,* are characteristic.   *Glyptodendron* of the Clinton is probably a Lycopodiaceous plant.

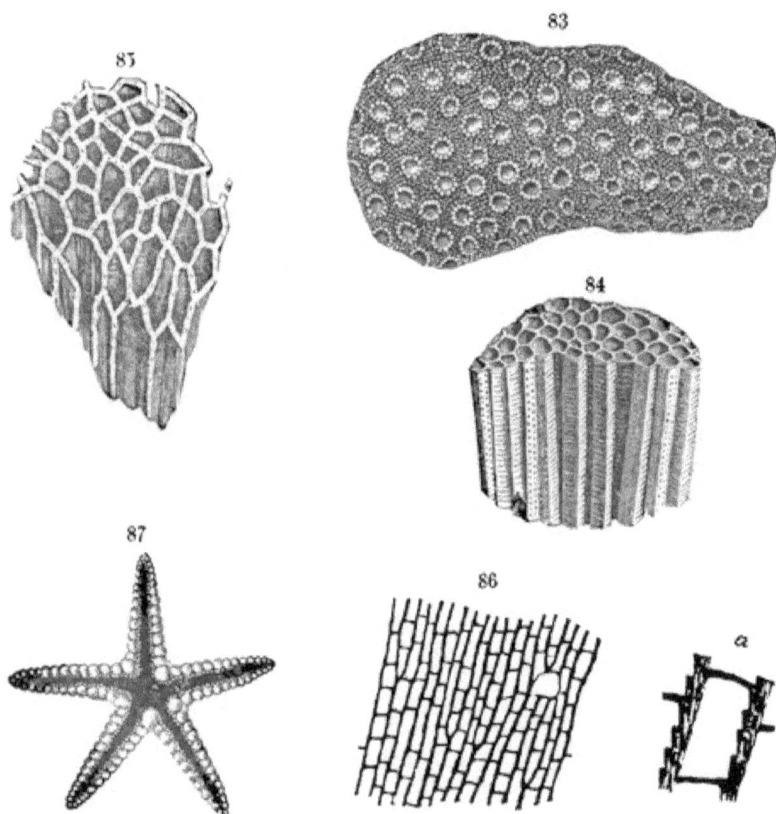

SILURIAN FOSSILS.

Fig. 83.—*Heliolites speciosus* 84. *Favosites Gothlandica.*   85. *Halysites catenulata.* 86. *Dictyonema Websteri.*   87. *Palæaster Niagarensis.*

3. *Salina Series.* In Canada—Shales, marls, dolomites and rock salt of Goderich in Ontario. This is a local series confined to the interior basin of North America, and marking a period of elevation and dry climate with deserts and salt lakes. The Guelph limestone and dolomite of Ontario is a transition deposit between this and the Niagara. *Megalomus Canadensis,* a large lamellibranchiate, is characteristic of the Guelph limestone. There are also species of *Stromatopora, Murchisonia, Cyclonema,* &c.

4. *Helderberg Series.* In Canada—Limestone of St. Helen's Island, Montreal; Upper Limestones of Anticosti; Cape Gaspe limestone; Upper Arisaig series, Nova Scotia. European equivalents—Ludlow Series of England; Etage F, G, of Bohemia.

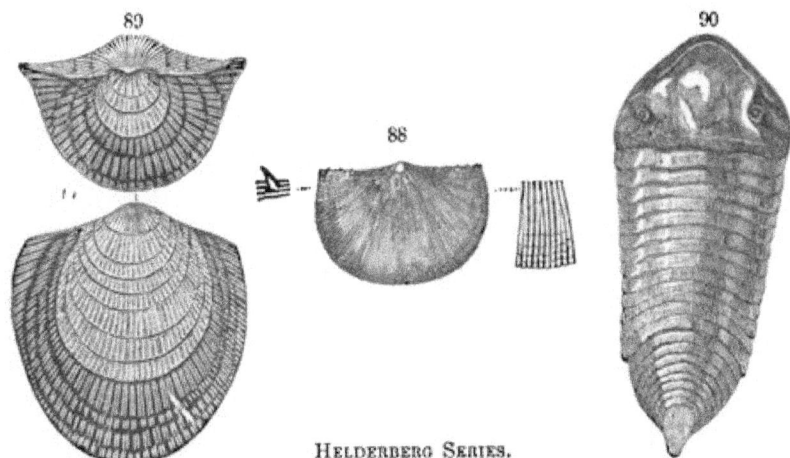

HELDERBERG SERIES.

Fig. 88—*Chonetes Nova Scotica.* 89. *Atrypa reticularis.*
90. *Homalonotus delphinocephalus.*

FOSSILS.—*Pentamerus galeatus, P. pseudo-galeatus, Rhynchonella ventricosa.* Species of *Merista, Chonetes, Eatonia, Stricklandinia, Tentaculites* and *Eurypterus,* are characteristic. The earliest known Scorpions and Insects occur in the Silurian. Fossil plants of genus *Psilophyton, Nematophyton,* &c., occur. Earliest fossil fishes—*Placoganoids* and *Selachians.* In America—*Pteraspis Acadica,* Matthew, from the Helderberg of New Brunswick, *Palæaspis bitruncata, P. Americana,* Claypole, Salina of Pennsylvania.

Distribution.—The Silurian rocks are well developed in the district extending north-westward from the Niagara river to Lake Huron. They occupy a large area in Quebec and Northern New Brunswick, extending S. W. from Gaspe and the Bay de Chaleur; and isolated areas occur in Nova Scotia and Southern New Brunswick.

In the East, where they have been involved in the folds extending
South-westwardly in the Apalachians, they have been much disturbed,
more or less altered and affected with slaty structure.   In the interior
plateau and in the area of the Gulf of St. Lawrence they remain flat or
inclined only at very small angles.   Extensive ejections of dolerite and
other igneous rocks occurred in the later part of the Silurian period.
Examples of these are afforded by the great sheets of dolerite interstratified with the Upper Silurian in Northern New Brunswick, (C. Bon
Ami, &c.) and in the trappean masses occurring in the valley of the St.
Lawrence, and to which Mount Royal and Belœil mountains belong.
The manner in which the dykes of these mountains cut the Utica beds
and the association of their agglomerates with the Helderberg limestones
show that these were active volcanoes at the close of the Silurian.

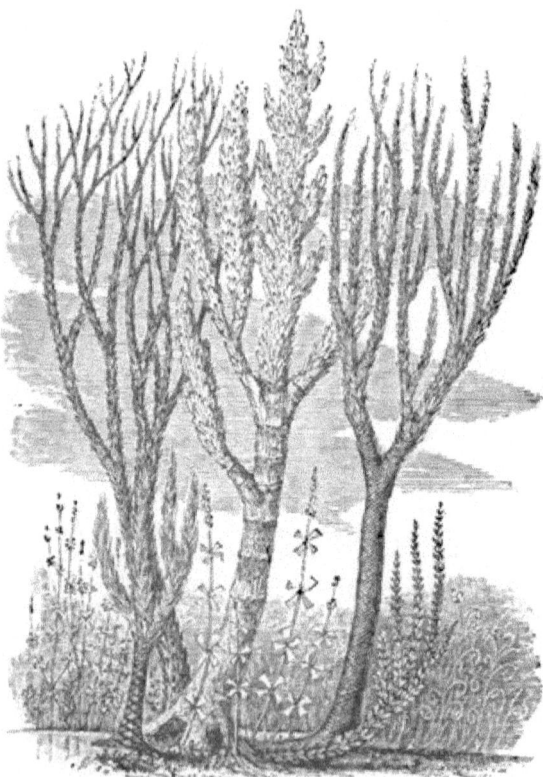

Fig. 91.—Silurian vegetation of N. America.   *Protannularia, Berwynia,*
Nematophyton, Arthrostigma, Psilophyton.

The earliest known land animals belong to the Silurian. In 1886 Dr. Lindström announced the discovery of a well-preserved specimen of a true scorpion, which he named *Palæophoneus nuncius*, in the Upper Silurian of Sweden; and in December of the same year a similar discovery in Scotland was announced by Dr. Hunter. In the following year, Prof. Whitfield of New York described and figured a third species in the Lower Helderberg series of the State of New York. Thus this form of life has been at one bound, and in three different localities, carried back from the Carboniferous to the Silurian, a remarkable instance of the nearly simultaneous discovery of new facts, in different places and by different observers.

The insects had previously been traced back to the Devonian or Erian period, and the scorpions would now have antedated them, but for another discovery made in Spain by M. Donville, and communicated to the Academy of Sciences by M. Charles Brongniart, in December, 1884. This is a wing of an insect in the sandstone of the Middle Silurian, probably equivalent to our Niagara series in Canada. This wing is shown by its venation to belong to the Blattidæ or cockroaches, a group already well known in the Carboniferous, where they seem to have thriven on the abundant vegetable matter of that period. It differs, however, in some of the details of venation from any living or fossil species known. Brongniart proposes for it the name *Protoblattina Donvillei*, and as the beds containing this insect are probably a little older than any of those containing the scorpions above referred to, this discovery makes the cockroaches, still so numerous and voracious a family of insects, the oldest known air breathing animals.

The Silurian is also characterized by the earliest known fishes, which belong to the groups of the Placoganoids or plate-bearing Ganoids, and the Selachians or sharks. These early fishes had in this period to contend for the mastery of the seas with gigantic Orthoceratites and with large and formidable Crustaceans, (Pterygotus, &c.)

On the land the few known fragments of plants indicate a meagre flora of Protogens and Acrogens.

Modiolopsis rhomboidea, Silurian.

Minerals.—The great Salt deposits of Goderich, &c., in Ontario, occur in the Salina series, which also affords Gypsum. The following section from the Geological Survey* illustrates these deposits :—

	Feet.	Inches.
Pleistocene Clays, &c........	76	9
Dolomite with Limestone bands........................	278	3
Limestone with Corals, Chert, and layers of Dolomite..	276	0
Dolomite....	243	0
Variegated Marls with layers of Dolomite............	121	6
Rock Salt...	30	11
Dolomite and Marl.............................. ...	32	1
Rock Salt.......................................	25	4
Dolomite........	6	10
Rock Salt......	34	10
Marls and Dolomite.....................	80	7
Rock Salt.............	15	5
Dolomite and Anhydrite......................... ..	7 ; 0	
Rock Salt...........	13	6
Marls.............	135	6
Rock Salt......................	6	0
Marls..	132	0
Total........ ....1572 feet.		

Limestones of valuable quality abound in the Niagara series, and roofing slates are afforded by some of the altered shales in some parts of Eastern Canada.

IV. ERIAN SYSTEM.

(Devonian of English Geologists.)

1. *Corniferous Series*—Corniferous limestone and associated sandstones in Ontario, Lower Gaspé sandstones. European equivalents—Plymouth and Linton groups of Devon ; Eifel limestones, spirifer sandstone of Germany ; old red sandstone of Scotland and West of England. The Oriskany sandstone, which lies at the base of the Corniferous series, is by some regarded as the upper member of the Silurian. In Canada, however, it is more closely connected with the Erian.

FOSSILS.—Placoganoid and Ganoid fishes abound. Abundant corals of genera *Favosites, Heliophyllum, Eridophyllum, Cystiphyllum, Zaphrentis,* &c. Plants— *Nematophyton Logani* and *Psilophyton princeps.*

* Sterry Hunt, 1876.

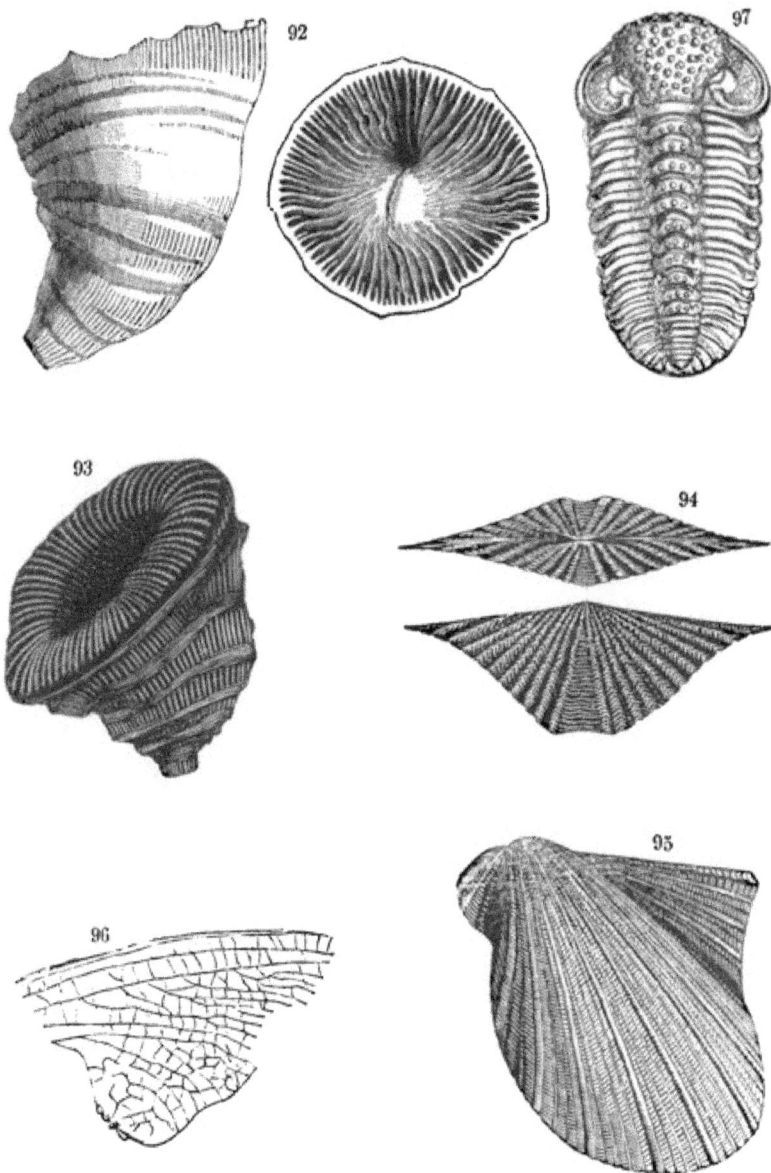

ERIAN OR DEVONIAN FOSSILS.

Fig. 92.—*Zaphrentis prolifica*. 93. *Heliophyllum Halli*. 94. *Spirifer mucronatus*.
95. *Pterinea flabella*. 96. *Platephemera antiqua*. 97. *Phaceps bufo*.

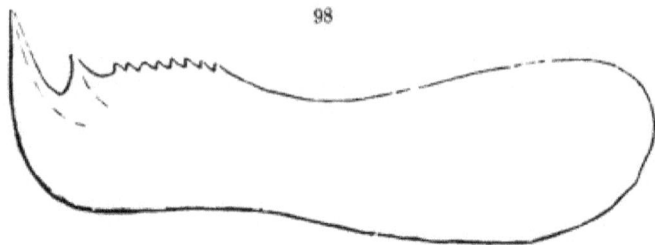

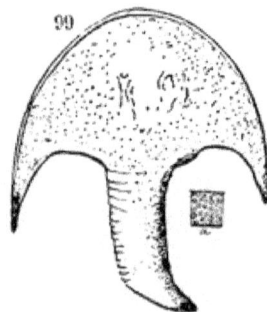

ERIAN FISHES.

Fig. 98.—*Jaw of Dinichthys* (reduced).   99. *Cephalaspis Dawsoni* (reduced).
(a) Sculpture.

2. *Hamilton Series.*—Hamilton shales of Western Ontario. Middle part of Gaspé sandstones, Cordaite shales of St. John, New Brunswick. European equivalents—Middle Devonian of England and Scotland; upper part of Eifel formation.

FOSSILS.—*Spirifer mucronatus* and *Atrypa reticularis* and *aspera* are common. The genus *Goniatites* appears. Fishes of genera *Dinichthys*. Trilobites of genus *Phacops*. Numerous fossil plants of the genera *Calamites, Lepidodendron, Psilophyton, Archæopteris, Cordaites,* &c. Several insects (*Platephemera*, &c.) appear in the St. John shales. Earliest Decapods (*Palæopalemon*).

3. *Chemung Series.* In Canada—Shales, &c., of Kettle Point, Lake Huron, Upper Gaspé sandstone, upper sandstone and conglomerate of St. John, New Brunswick. European equivalents—Upper old red sandstone of Scotland, Kiltorcan beds in Ireland, Petherwin group of Devon, Cypridina shale of Germany.

FOSSILS.—Many Lamellibranchiates of genera *Pteronites, Avicula,* &c. Fishes of genera *Holoptychius, Pterichthys,* &c. Peach has discovered Millipedes of two species in the Erian of Scotland. Ferns of genera *Archæopteris, Cyclopterie,* &c.

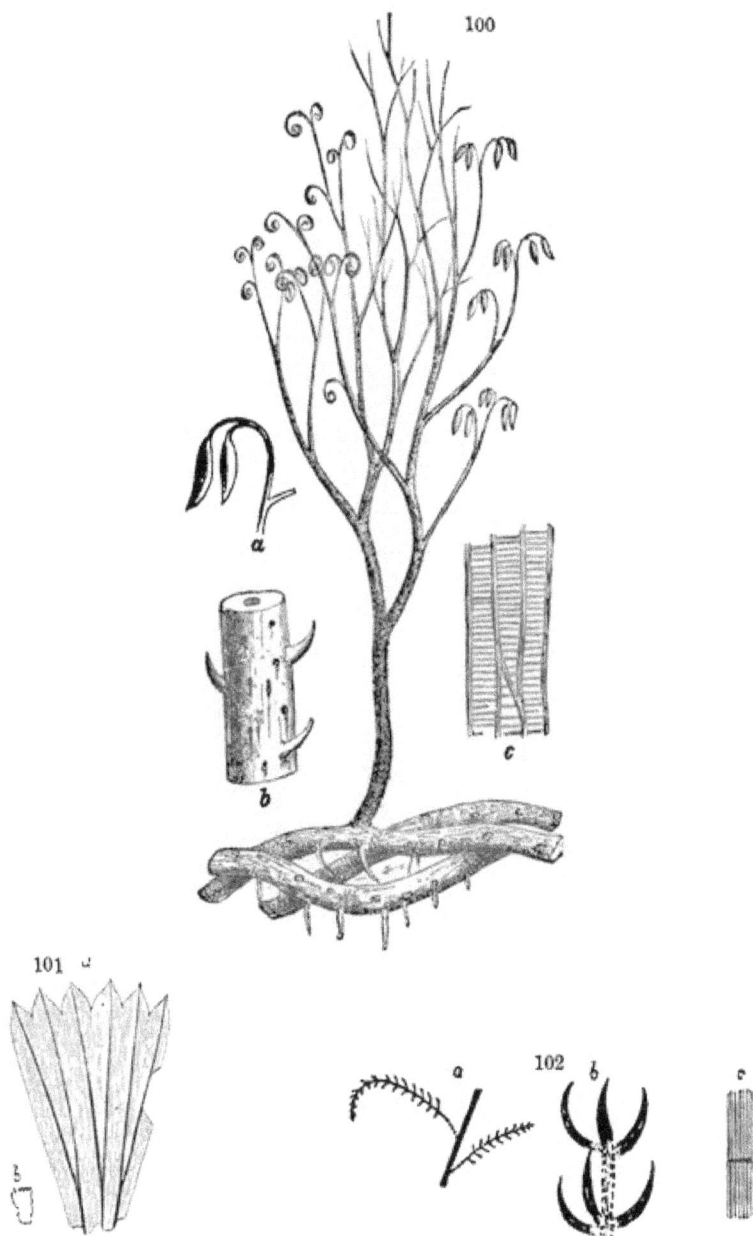

ERIAN OR DEVONIAN PLANTS.

Fig. 100.—*Psilophyton princeps. a*, fruit, *b*, stem enlarged, *c*, scalariform vessels of axis. 101. *Sphenophyllum antiquum, a*, magnified, *b*, natural size. 102. *Asterophyllites parvula, a*, natural size, *b, c*, portions magnified.

Distribution. These rocks occupy the peninsula of Ontario between Lakes Erie and Huron. They occur largely in the region south of Lake Erie and elsewhere in the United States. They are extensively developed in Gaspé and the Bay de Chaleur and also in Southern New Brunswick. In the maritime regions however, last mentioned, the great limestones, so rich in corals in Ontario, are wanting, and the whole system is represented by shallow-water beds, while fossil plants and remains of fishes prevail to the exclusion of strictly marine forms. The following table shows the comparative development of the system on the Atlantic margin and the Continental Plateau :—

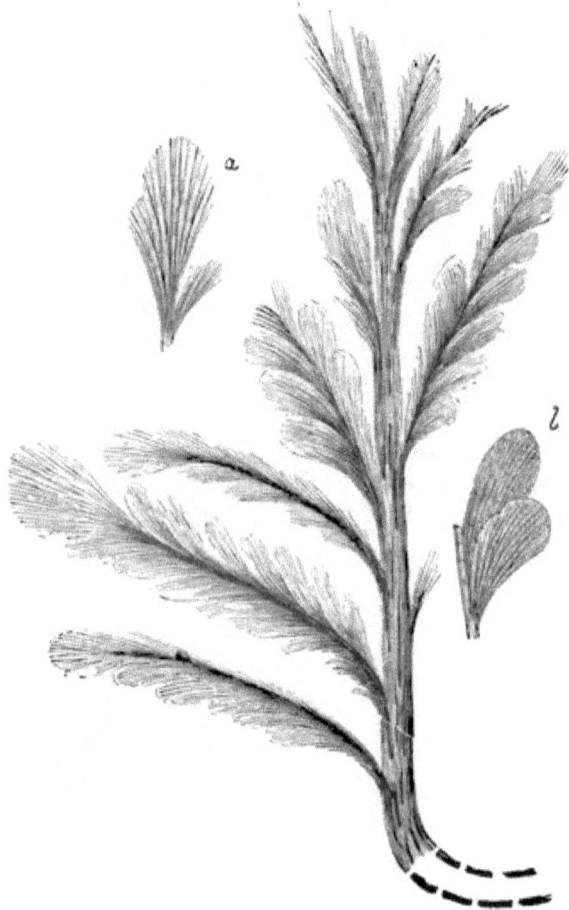

Fig. 103.—*Archæopteris Jacksoni, a, b,* portions showing venation.

Devonian, or Erian of America.

SUBDIVISIONS.	NEW YORK AND WESTERN CANADA.	GASPÉ AND BAY DE CHALEUR.	SOUTHERN NEW BRUNS. WICK AND NOVA SCOTIA.
Upper Devonian or Erian.	Chemung Group.	Upper Sandstones. Long Cove, Scaumenac.	Mispec Group. Shale, Sandstone, and Conglomerate. Sandstones near Middle R., Pictou ?
Middle Devonian or Erian.	Hamilton Group.	Middle Sandstones. Bois Brulé, Cape Oiseau, etc.	Little R. Group (including Cordaite shales and Dadoxylon Sandstone).
Lower Devonian or Erian.	Corniferous and Oriskany groups*	Lower Sandstones. Gaspé Basin. Little Gaspé Campbellton.	Lower Conglomerates, etc. Nictaux and Bear River Series (Oriskany).

A similar difference obtains between the Atlantic margin and the interior continental area in Europe, as evidenced by the comparison of the old Red Sandstone of Scotland with the Eifel limestones in Germany.

The Erian is emphatically the period of the reign of fishes, when animals of that class first became dominant in the waters. The rich fish fauna of the Devonian of Scotland has long been known, and the discovery of *Cephalaspis* and *Machaeracanthus* in the Gaspé sandstones, by the author, in 1869,* showed that both Placoganoid and Selachian fishes existed in Canada in the Lower and Middle Devonian. More recently Mr. Whiteaves has described, from the Lower Devonian of Campbellton, fishes of the genera *Coccosteus*, *Cephalaspis* and *Ctenacanthus*, and from Upper Devonian beds at Scaumenac Bay, other species of the genera *Pterichthys*, *Diplocanthus*, *Phaneropleuron*, *Glyptolepis*, *Cheirolepis*, and of a new genus named by him *Eusthenopteron*.† The evidence of the relative ages of the beds as indicated by the fishes corresponds perfectly with that deduced by the writer from the fossil plants which accompany them.

The Erian is also the age in which we first find, in the Palæozoic, the evidence of extensive forests and of great vegetable growth. We have remains of Land plants, as already stated, in the Silurian. But in the Erian we find a profusion of new forms, introduced however not in the beginning of the period but more especially in its middle portion. The Lower Erian flora is meagre, and the prevalent forms are *Psilophyton*,

* Geological Magazine, 1870.

† Gesner first noticed the Scaumenac fishes which were subsequently collected by Ells and Foord.

Arthrostigma and *Nematophyton.* In the middle Erian we have Gymnosperms, represented by *Dadoxylon, Cordaites,* and the fruits known as *Antholithes* and *Cardiocarpum.* Of Acrogens we have *Lepidodendron* and *Leptophleum* and *Lycopodites,* representing the Lycopods, *Calamites,* representing the Equisetaceæ and numerous genera and species of ferns, both tree-ferns and herbaceous species. We have also a vast exuberance of the humble aquatic plants known as Rhizocarps, their sporocarps filling some thick beds of shale in such a manner as to render it highly bituminous.

Mineral Products.—The Erian is especially remarkable as a source of petroleum, which seems to abound in the Corniferous and Hamilton groups in certain districts in Ontario and in the North-West.

V. CARBONIFEROUS SYSTEM.

1. *Horton Series.* Lower Carboniferous Shales and Conglomerates, Horton Bluff, &c., in Nova Scotia. Equivalents in United States—Vespertine group of Pennsylvania ; Waverly sandstone (in part), Ohio ; Kinderhook and Marshall groups of Illinois and Michigan ; lower or false coal measures of Virginia. European equivalents—Tweedian group or Calciferous sandstones of Scotland ; Carboniferous shale and Coomhala grits of Ireland ; Culm formation of Germany, Graywacke of Vosges.

FOSSILS.—Fishes of genera *Rhadinichthys, Rhizodus, Acrolepis, Ctenacanthus,* &c. Footprints of earliest known Batrachians: *Lepidodendron corrugatum, Aneimites, Acadica, Cordaites,* &c.

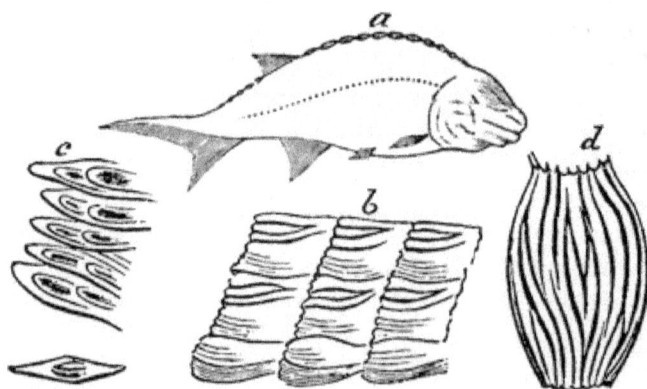

Fig. 104.—*Palæoniscus modulus,* Dn. *a,* Outline, natural size. *b,* Series of Scales enlarged, seen from inside. The lower row are those on mesial line. *c,* Surface of exposed part of scale from side and from upper lobe of tail, showing sculpture enlarged. *d,* One of the dorsal scales, enlarged.

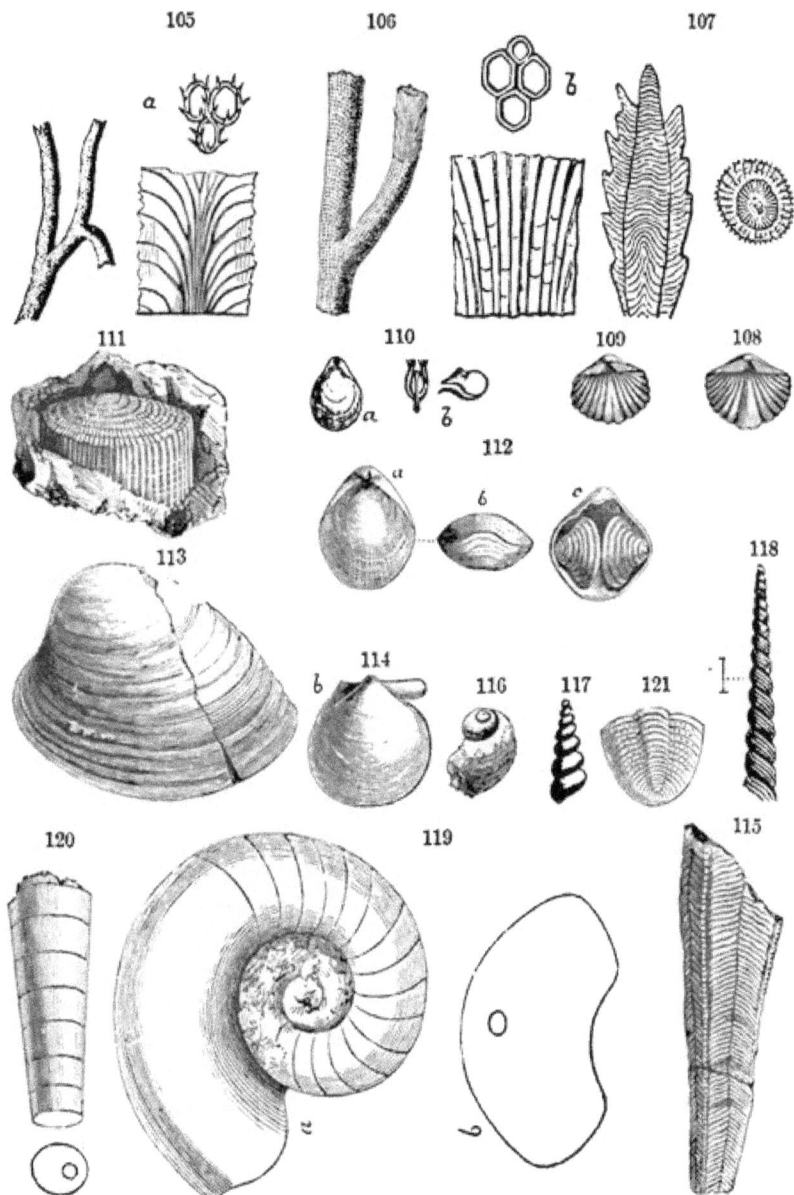

LOWER CARBONIFEROUS FOSSILS.

Fig. 105.—*Stenopora exilis.* 106. *Chætetes tumida.* 107. *Lithostrotion Pictoense.* 108. *Spirifer acuicosta.* 109. *Spirifer cristata.* 110. *Centronella anna.* 111. *Productus semireticulatus.* 112. *Athyris subtilita.* 113. *Cardiomorpha Vindobonensis.* 114. *Aviculopecten simplex.* 115. *Conularia quadrisulcata.* 116. *Naticopsis dispassa.* 117. *Murchisonia gypsea.* 118. *Loxonema acutula.* 119. *Nautilus avonensis.* 120. *Orthoceras vindobonense.* 121. *Phillipsia Howi.*

2. *Windsor Series.* In Canada—Lower Carboniferous limestones and gypsiferous series of Nova Scotia and New Brunswick. Equivalents in United States—Burlington, Keokuk and Chester limestones of Illinois. European equivalents—Old Mountain or Carboniferous limestone of England ; Calcaire Condrusien of France ; Kohlen-kalkstein of Germany ; Fusulina limestone of Russia.

FOSSILS.—Marine Invertebrates of genera *Fusulina, Lithostrotion, Cyathophyllum, Fenestella, Productus, Terebratula, Athyris, Spirifer, Aviculopecten, Macrodon, Conularia, Nautilus, Orthoceras, Phillipsia,* &c.

3. *Millstone grit.* Canadian types—Sandstones and conglomerates between the Carboniferous limestones and the coal formation, in Nova Scotia and New Brunswick. In United States—Seral conglomerate of Pennsylvania, Lower Carboniferous sandstone of Kentucky, Alabama and Virginia, Chester group of Illinois in part. European equivalents—Millstone grit and Yoredale rocks of England ; Moor rock of Scotland ; Jungste Grauwacke of the Hartz, Saxony and Silesia.

FOSSILS.—Plants similar to those of the Coal formation.

4. *Coal Formation.* In Canada—Productive coal measures of Nova Scotia and New Brunswick. In United States—coal formation of Pennsylvania, Ohio, Illinois and Michigan, represented in the west by marine limestones, &c. In Europe—the coal formations of Scotland, England, France, Germany, &c.

FOSSILS.—Land plants of genera *Araucaroxylon, Sigillaria, Lepidodendron* and *Calamites,* and *Ferns* and allied plants. Fishes of genera *Pulæoniscus, Rhizodus, Diplodus, Gyracanthus,* &c. Batrachians of genera *Baphetes, Dendrerpeton, Hy'onomus, Anthracosaurus,* &c. Insects, Millepedes, Arachnidans and Decapod Crustaceans.

ORIGIN OF COAL.

All ordinary beds of coal consist of compressed and carbonized vegetable matter, principally the cortical and other more durable and least permeable tissues of plants. That this matter has accumulated *in situ* on swampy flats and depressions, is shown by the occurrence of fossil soils or "underclays" full of roots, below the coal-seams, by the vegetable debris spread out in the shales which overlie the coal, and by the frequent occurrence in these "roof-shales" of erect trees rooted on the surface of the coal. (See Figures, page 92).

The chemical relation of vegetable matter to ordinary bituminous coal is seen in the following table :—

Cellulose, - - - -	$C_{24} H_{20}$	O_{20}
Cork, - - - - -	$C_{24} H_{18\frac{2}{10}}$	$O_{6\frac{7}{10}}$
Bituminous Coal (Regnault),	$C_{24} H_{10}$	$O_{3\frac{3}{10}}$

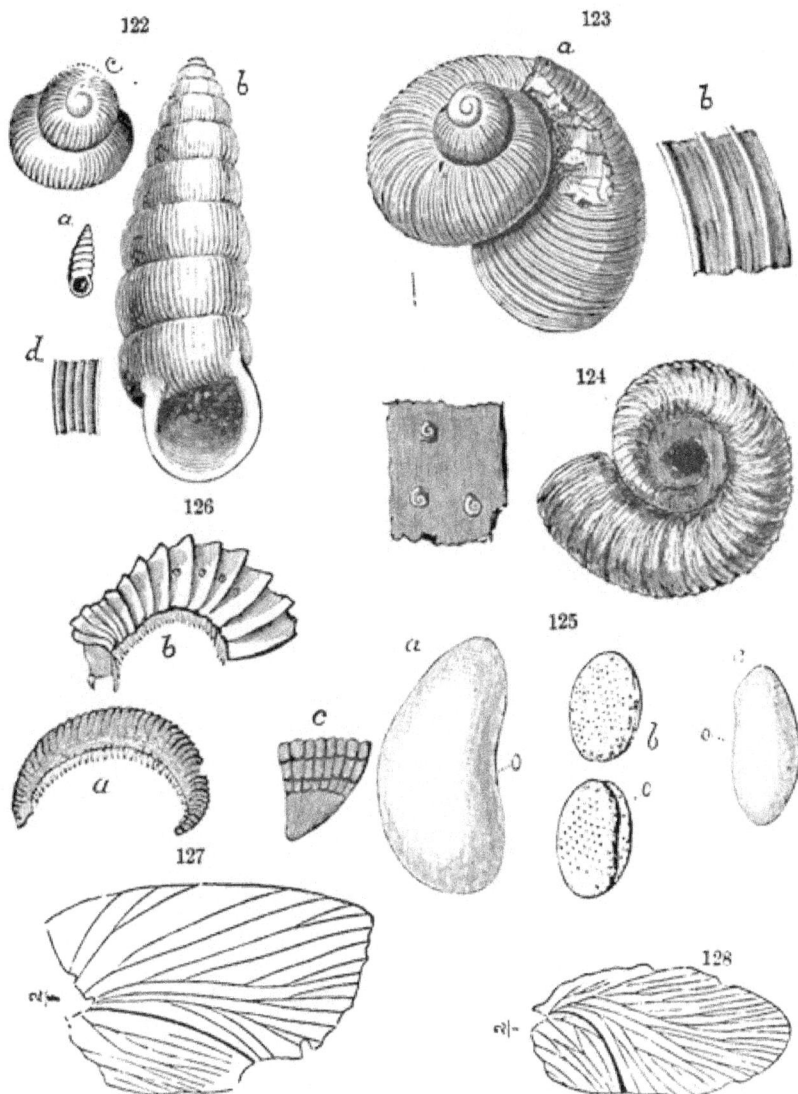

COAL-FORMATION FOSSILS

Fig. 122. *Pupa vetusta;* a, natural size, b, enlarged, c, apex, d, sculpture. 123. *Conulus priscus;* a, enlarged, b, sculpture. 124. *Spirorbis carbonarius.* 125. *Entomostracans;* a, *Carbonia bairdeoides,* Jones; b, *Carbonia inflata,* Jones; c, *Cythere.* 126. *Millipedes;* a, *Xylobius, sigillariæ;* b, *Archiulus Xylobiodes;* c, *Xylobius farctus.* 127. *Blattina Bretonensis.* 128. *Blattina Heeri.*

Distribution.—In Canada the Carboniferous occupies considerable areas in Nova Scotia and New Brunswick, and includes the extensive and valuable coal fields of Cumberland, Pictou and Cape Breton. The Carboniferous district of New Brunswick while extensive in area is as yet known to contain only thin beds of coal. The thickest beds occur in the vicinity of the ancient shores of the older formations in Pictou, Cape Breton and Southern Cumberland. These coal beds from their resting on fossil swamp soils or underclays and being covered with beds holding erect trunks of trees, are known to have been produced in situ, and to represent buried bogs of the period. The microscope shows that the coal itself is made up of layers of vegetable debris, largely cortical and epidermal tissues of trees, and sporocarps and other remains of fructification. Woody matter proper appears in the form of mineral charcoal.

In the Rocky Mountains and in British Columbia, Carboniferous rocks are largely developed, but consist principally of quartzites and limestones associated with igneous rocks. In British Columbia the limestones contain foraminifera of the genera *Fusulina and Loftusia.*

The forests of the Coal period were remarkable for the great abundance of Sigillariæ, and the Stigmaria roots of these trees are present in nearly all the underclays. Conifers, (*Dadoxylon*), *Cordaites*, Ferns, *Equiseta* and *Lepidodendra* are also extremely abundant. The flora throughout consists of Gymnosperms and Acrogens, but the species differ somewhat in the different members of the group.

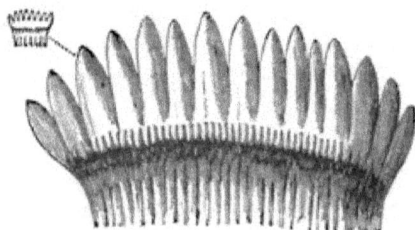

Fig. 129.—Tooth of *Ctenoptychius cristatus*, N.S. ; natural size and magnified.

In addition to insects and scorpions we now have on the land many myriapods and land snails, and above all, numerous batrachians, some of which, those of the Labyrinthodent and Microsaurian groups, presented characters in advance of those of the class in modern times. The genera *Eosaurus, Baphetes, Dendrerpeton* and *Hylonomus,* may be mentioned as examples. (Fig. 130.)

Useful Minerals.—The Lower Carboniferous of Nova Scotia abounds in Limestone and Gypsum. It also contains ores of Iron (Limonite and Suderite) and of Manganese (Pyrolusite). Valuable sandstones for building, and grindstones are also quarried in the Millstone Grit and Coal formation. This formation is however most remarkable for the great workable coal beds, now mined in various places, and the produce of which is extensively exported from Nova Scotia and Cape Breton, more especially from the Pictou and Springhill mines and from the various collieries in Eastern Cape Breton.

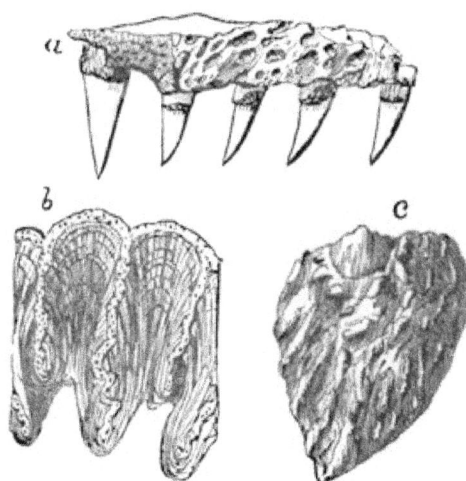

Fig. 130.—*Baphetes planiceps*, Owen. *a*, Fragment of Maxillary bone, showing sculpture, four outer teeth, and one inner tooth; natural size. *b*, Section of inner tooth; magnified. *c*, Dermal scale; natural size.

The Permo-Carboniferous occupies some space in the south of Prince Edward Island; and the Lower Carboniferous, locally termed the Bonaventure formation, extends into the east of Quebec. A limited area, including beds of coal, occurs in western Newfoundland. In the west, rocks of Carboniferous age occur in the Rocky Mountains and in British Columbia, but, so far as known, without beds of coal.

The following section from Acadian Geology represents a bed of coal with its accompaniments:—(Fig. 131.)

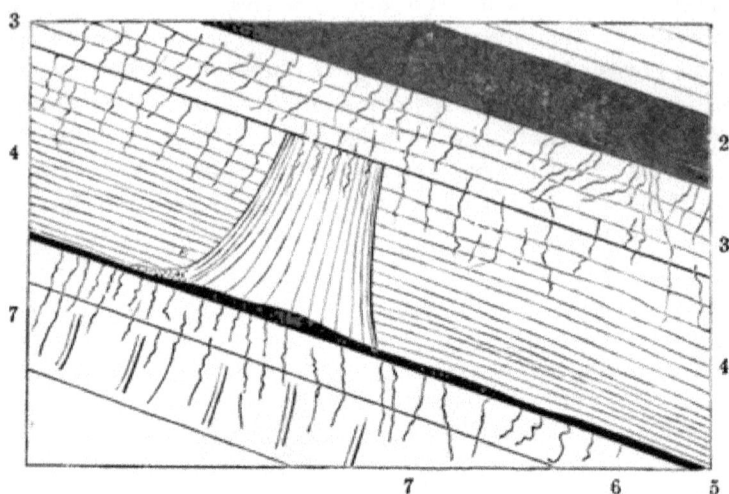

Fig. 131.—Section from the Coal formation of S. Joggins. 1. Shale. 2. Shaly coal,
1 foot. 3. Underclay with rootlets, 1 foot 2 inches. 4. Gray sandstone
passing downwards into shale, 3 feet.   Erect tree with Stigmaria roots *(e)*
on the coal. 5. Coal, 1 inch. 6. Underclay with roots, 10 inches. 7. Gray
sandstone, 1 foot 5 inches. Stigmaria rootlets continued from the bed
above; erect *Calamites.* 8. Gray shale, with pyrites.—flattened plants.

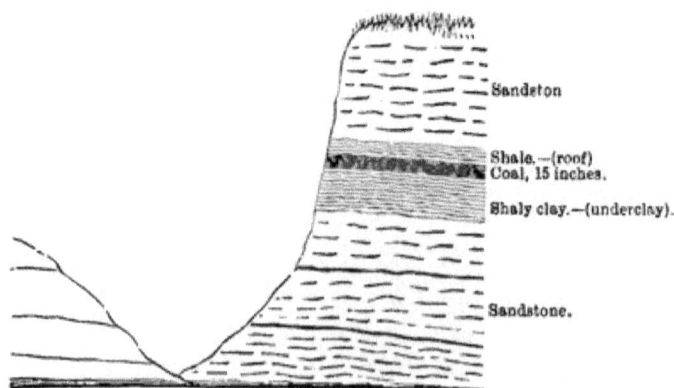

Fig. 131a.—Section on Coal Creek near Richebucto.—Dr. Robb.

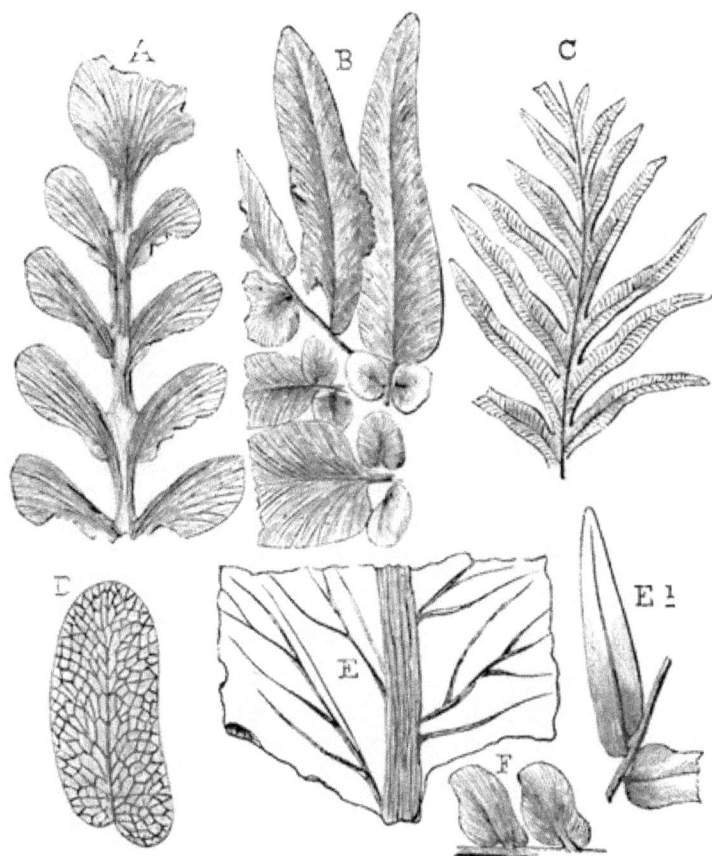

FIG. 132.—CARBONIFEROUS FERNS.

A   *Odontopteris subcuneata* (after Bunbury)
B   *Neuropteris cordata.*
C   *Alethopteris lonchitica.*
D   *Dictyopteris obliqua* (after Bunbury).
E   *Phyllopteris antiqua*, mag.   (E¹) nat. size.
F   *Neuropteris cyclopteroides.*

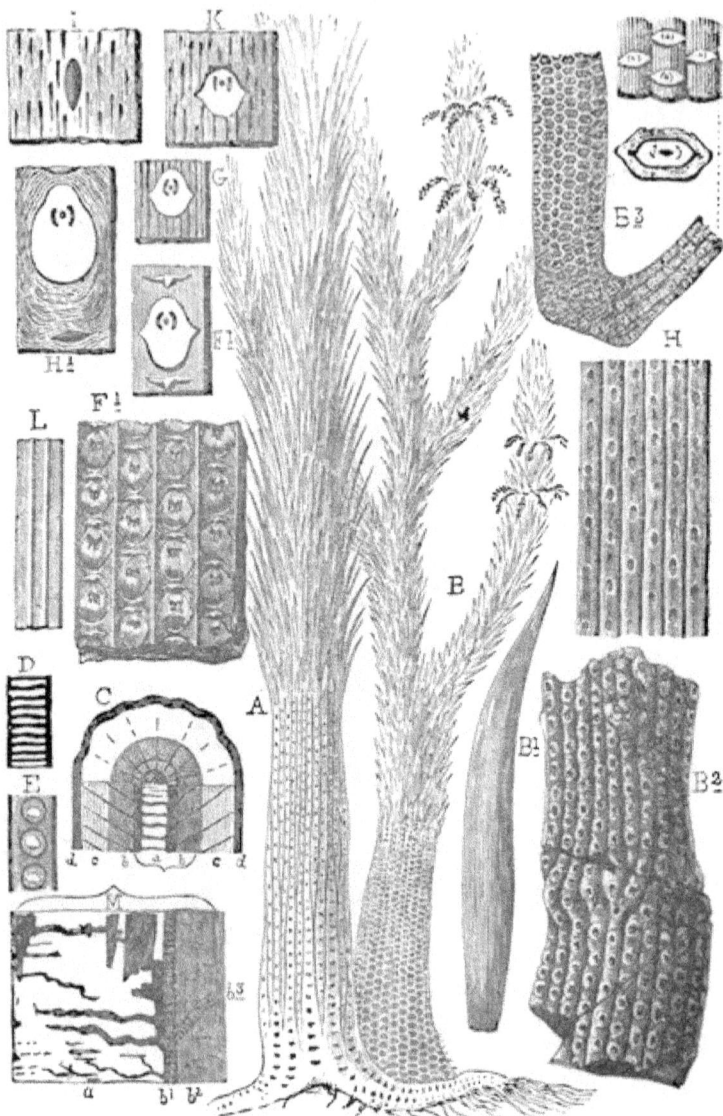

FIG. 133.—CARBONIFEROUS SIGILLARIÆ.

A *Sigillaria Brownii*, restored. B *S. elegans*, restored. B1 Leaf of *S. elegans*. B2
Portion of decorticated stem, showing one of the transverse bands of fruit-scars.
B3 Portion of stem and branches, reduced—and scars, nat. size. C Cross section of *S. Brownii* (?), reduced, and portion at (M), nat. size. (*a*) Sternbergia
pith, (*b1*) Scalariform vessels, (*b2*) Discigerous cells, (*c*) Inner bark, (*d*) Outer
bark. D, E, Tissues mag. F *Sigillaria Bretonensis*, ⅔. G *S. striata*. H *S.
eminens*, reduced. I *S. catenoides*. K *S. planicosta*. L Leaf.

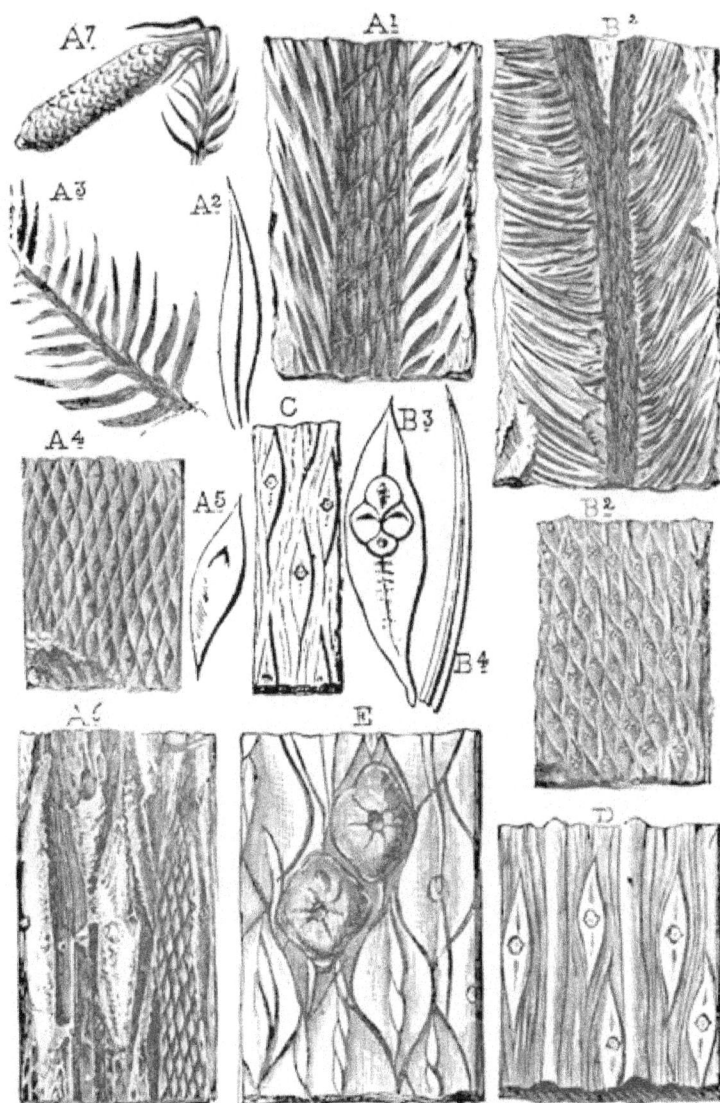

FIG. 134.—CARBONIFEROUS LEPIDODENDRA.

A   Branch and leaves of *L. Pictoense*, $\frac{2}{3}$ nat. size.   A2 Leaf.   A3 Twig and leaves, $\frac{2}{3}$.   A4 Portion of bark, $\frac{2}{3}$.   A5 Leaf-scar.   A6 Bark of old stem furrowed by growth, $\frac{2}{3}$.   A7 Cone, $\frac{2}{3}$.

B1   *L. personatum*, leafy branch, $\frac{2}{3}$.   B2 Portion of bark, $\frac{2}{3}$.   B3 Areole enlarged.   B4 Leaf.

C   *L. plicatum*, bark of old stem.

D   *L. rimosum*, old stem with furrows, $\frac{2}{3}$.

E.   *L. undulatum*, showing furrows and scars of cones, $\frac{2}{3}$.

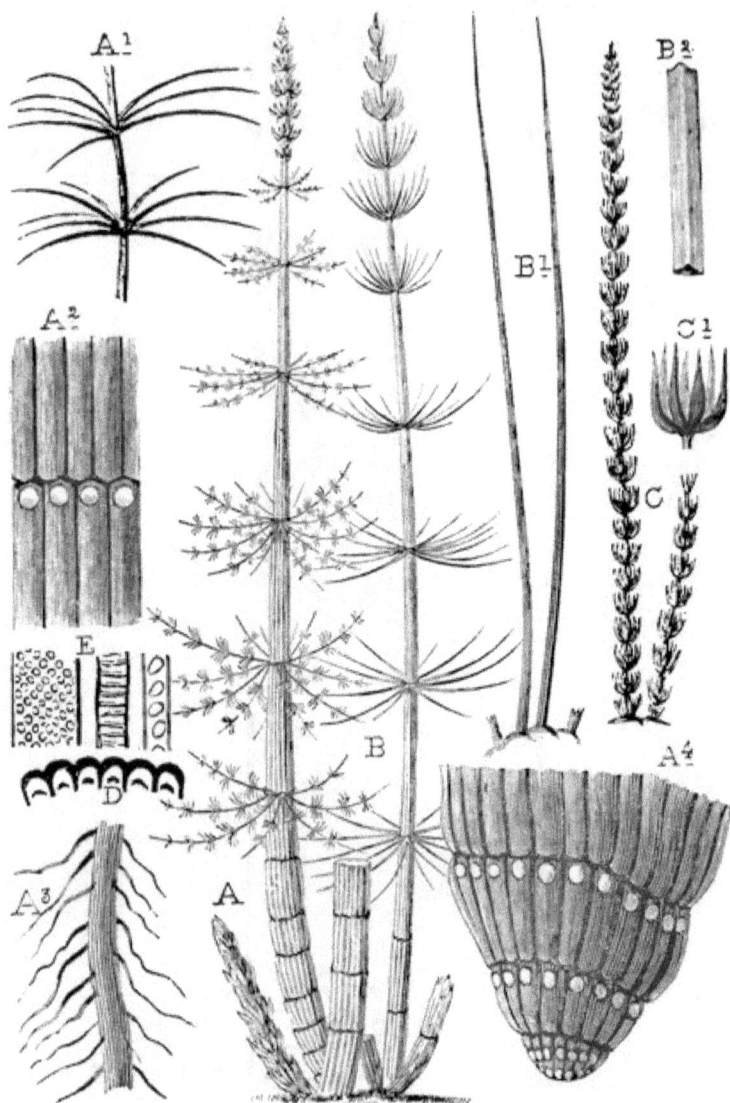

FIG. 135.—CARBONIFEROUS CALAMITES.

A   *Calamites Suck v* restored.   B1 Leaves.
· A1 Foliage.   B2 Leaf enlarged.
A2 Ribs and Scars.   C Leaves of *C. nodosus.*
A3 Roots.   C1 Whorl, enlarged.
A4 Base of stem.   D Structure of stem.
B   *Calamites Cistii*, restored.   E Vessels, magnified.

VI. Permian System.

1. *Lower Permian.* Canadian type—Permo-carboniferous red sand-stones of Prince Edward Island and Eastern Nova Scotia. In United States—Permian sandstones of Virginia and limestones of Kansas and Nebraska. Upper Carboniferous beds of Illinois, holding remains of reptiles. European equivalents—Lower Permian sandstones of England. Rothelicgendes of Germany, Lower Permian Sandstones and Limestones of Russia.

Fossils.—For the most part generically similar to those of the Carboniferous. The earliest true reptiles appear.

2. *Upper Permian.* Not represented in Canada, but marine lime-stones of this age occur in Kansas and westward. In England it is rep-resented by the important formations of the Marl slate and Magnesian limestone ; in Germany by the copper slate and zechstein ; and in Russia by the copper sandstones and gypsiferous limestone.

Fossils.—Reptiles of genus *Proterosaurus.* Fishes of genus *Palæoniscus.* Mol-lusks of genera *Pseudomonotis, Myalina, Productus, Fenestella,* &c.

The Permian or Permo-carboniferous of Prince Edward Island does not as yet admit of any division into distinct groups, and it rests conform-ably on the Upper Coal-formation without any stratigraphical break. It is characterised, as will be seen in the section following, by a prevalence of sandstones and shales coloured by the red oxide of iron.

Section on the south coast of Prince Edward Island, at Gallas Point.

Feet.

1. Brown and reddish sandstones, with thin bands of concretionary limestones and calcareous conglomerate, with fragments of red clay or shale.......... 30

2. Red and mottled clay ; thin-bedded and alternating with argillaceous sandstones 39

3. Soft reddish sandstones...... 75

4. Brown sandstones with gray bands and layers of concretionary limestone, silicified and carbonized trees, Calamites and comminuted plants. 81

5. Concealed by marsh, probably clay and soft sandstone... 90

6. Red and gray sandstone with beds of red and mottled clay................. 60

7. Red sandstone with gray and white bands.............. 93

8. Reddish sandstone and gray bands, with nodules of oxide of iron, and Calamites 102

576

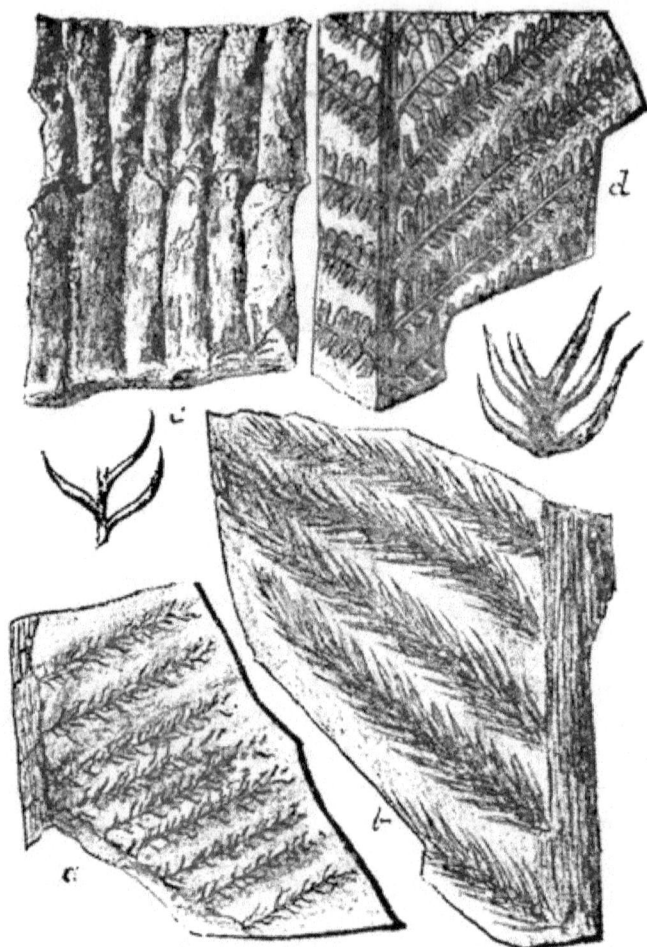

FIG. 136.   PLANTS OF THE PERMO-CARBONIFEROUS.
(Prince Edward Island.)

(a) Walchia gracilis. (b) W. robusta. (c) Calamites gigas. (d) Pecopteris arborescens.

General note on the Palæozoic.

It will be observed that each of the great systems of the Palæozoic has fossiliferous limestones indicating oceanic conditions in its central part, and beds indicating littoral conditions or shallow water above and below.   Thus each system constitutes a triplet of formations, and shows a gradual subsidence of the continental plateau followed by re-elevation, or, in other words, sandstones and conglomerates in the basal and upper part, limestone and shale in the middle. It was in the warm-water basins covering the plateau in times of subsidence that the successive marine faunas flourished. In intervening times the continental areas became elevated and extended, sometimes, as in the Carboniferous, in wide swampy flats; and forests and morasses spread themselves, while subærial denudation produced vast quantities of sand and gravel. These successive pulsations of the continental areas produced by successive collapses of the crust, constituted the physical environment of the flora and fauna of the Palæozoic. Finally a great lateral pressure and continental upheaval, sliding and crumpling, produced the Permian continents, and brought the Palæozoic to a close.

In connection with this it may be observed—(1) That the Carboniferous rocks share in the great foldings which have affected the older rocks in the Apalachian region, and in Nova Scotia and New Brunswick. (2) That the red sandstones of the Trias are deposited in troughs produced by the folding of the Carboniferous, and rest unconformably on the edges of the latter. (3) That the great trappean ejections of the Triassic age may be regarded as consequent on the preceding earth movements. (4) That the aqueous deposits of the Upper Permian and Lower Trias resemble each other in the prevalence of red sandstones, indicating that both belong to a period of transition. (5) That, throughout the Northern Hemisphere, the Palæozoic forms of life disappear almost entirely in the Permian and are succeeded by new types in the Trias.

In Canada the principal igneous ejections of the Palæozoic occurred in the Kewenian, (Lake Superior, &c.) in the Cambrian, (Rocky Mountains); in the Siluro-Cambrian (Nova Scotia, &c.); in the Silurian, (New Brunswick and valley of Lower St. Lawrence); in the Devonian, (granites of Nova Scotia and Quebec, Felsites, &c.); and in the Lower Carboniferous. (Nova Scotia and New Brunswick).

MESOZOIC PERIOD.

1. TRIASSIC SYSTEM.

1. *Bunter Sandstone.* In Canada—Lower new red sandstone of the Bay of Fundy and Prince Edward Island, associated with trappean rocks. In United States—Lower red sandstones of Connecticut and New Jersey. In the West, red and magnesian limestones overlying Carboniferous of Rocky Mountains. In Europe—Bunter sandstone of Germany, Lower Triassic red sandstones of England.

FOSSILS.—Conifers and Cycads. Footprints of Dinosaurs.

2. *Muschelkalk.* A marine limestone found in Germany and Eastern France, but not represented in England or Eastern America. In British Columbia and the Western United States the Triassic sandstones and slates with volcanic rocks, and the *Monotis* shales, may be partly of this age, and it may also be represented in the East by part of the Triassic coal formation of Virginia and South Carolina.

FOSSILS (in Europe)—*Encrinus moniliformis, Avicula socialis, Ceratites nodosus, Pemphyx Sueri,* &c. Fishes—*Hybodus,* &c. Reptiles, *Nothosaurus,* &c.

Fig. 137.—Jaw of Dormatherium Sylvestre—Trias.

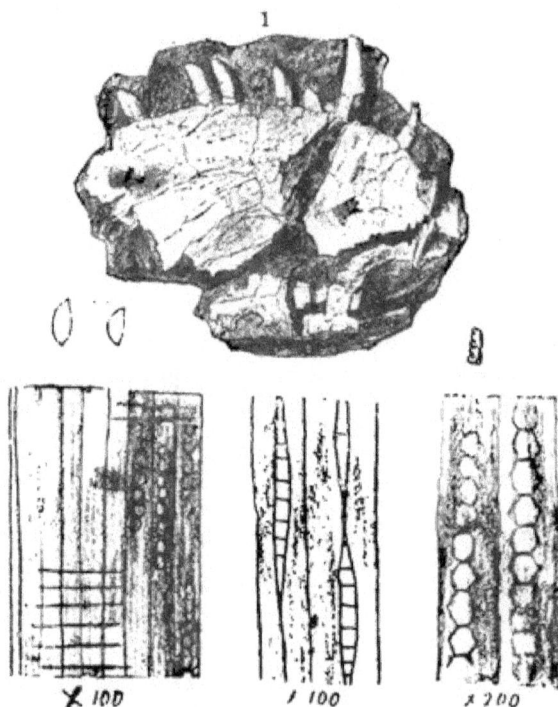

Fig. 138.—TRIASSIC FOSSILS.
(Prince Edward Island.)

1.—*Bathygnathus borealis* (Lower jaw), reduced. 2. *Araucaroxylon Edvardianum* (Structures magnified).

3. *Keuper Sandstone.* In Canada—Upper Triassic sandstones of Prince Edward Island and Bay of Fundy, and probably portions of the Trias of British Columbia. In United States—Upper red sandstone of North Carolina, &c. To the Upper Triassic are also usually referred the Mesozoic coal beds of Virginia and North Carolina with their associated sandstones and shales. In Europe—Saliferous series of England ; Keuper formation of Germany.

FOSSILS.—Plants, *Equisetum, Pterophyllum,* &c. Reptiles, &c., *Bathygnathus borealis,* footprints of Dinosaurs, *Labyrinthodon giganteum.* The earliest Marsupial mammals (*Microlestes, Dromatherium*).

Distribution of the Triassic in Canada.—This formation occupies a large part of Prince Edward Island and the basin of the Bay of Fundy, where its trappean beds form the "North Mountain" of Cornwallis and Annapolis. Rocks of this age also appear in the Rocky Mountains, in British Columbia and the Queen Charlotte Islands ; but in these Western regions their mineral character is very different from that which they present in the East.

Both in Nova Scotia and in New England the Triassic age was remarkable for the deposition of Red Sandstone in shallow bays and straits, and for the ejection of great beds and dykes of basaltic and amygdaloidal basic volcanic ash. On the coasts of the Bay of Fundy and Minas Basin the tufaceous beds are rich in zeolitic minerals of great beauty.

The sections show the characters of this formation in Prince Edward Island and in Western Nova Scotia. In the former Province, owing to the slight dips of the Permian and Triassic and their mineral similarity, it has proved difficult to define their boundaries ; but the Trias appears to rest in slight troughs of the Permian and to be partly composed of its debris.

Section in Orwell Bay, in ascending order, the beds resting on the Permian sandstones, &c., of Gallas Point, referred to under the Permian.

1. Bright red sandstones with white bands. 30
2. Red shales with white stains and red sandstones with cylindrical casts and fucoids 60
3. Red and purplish sandstones with gray bands and layers of ferruginous conglomerate with obscure remains of plants 88
4. Beach, probably representing soft beds 48
5. Red flaggy sandstone with conglomerate and concretions of red oxide of iron, containing remains of plants 50
6. Bright red sandstones and red shale with greenish stains. 30
7. Marsh, probably soft beds 24
8. Red Shale and green bands capped with bright red sandstones.... 75
                                                                          ———
                                                                          405

Fig. 139.—Section from Horton to Cape Blomidon, showing Trias, &c. a, Slates. (Silurian.) b, Carboniferous. c, Trias. d, Trap.

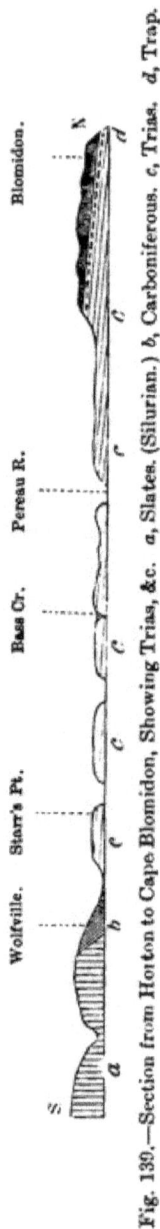

Fig. 140.—Triassic Trap and Tufa, Partridge Island, Nova Scotia. b, Carboniferous. c, Trias. c1, Tufaceous Trap. c2, Basaltic Trap. d, Modern Gravel.

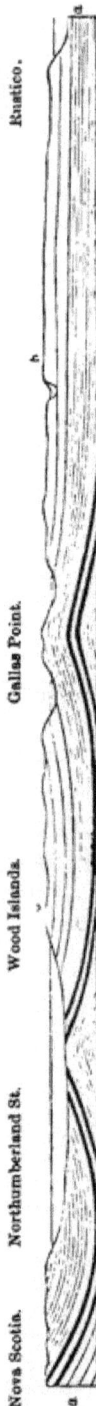

Fig. 141.—Triassic Trap in Coast Section East of Swan Creek. a, Shales and Sandstone of Carboniferous System, dip. E. N. E. b, New Red Sandstone, dips to S. W. c, New Red Sandstone with fragments of Trap. d, Trap Conglomerate and Breccia. e, Amygdaloidal Trap. f, Crystalline Trap.

Fig. 142.—Ideal Section, showing the probable relations of the Trias of Prince Edward Island to the Carboniferous of Nova Scotia. a, Carboniferous. b, Trias.

(Here the section is broken by Orwell Bay, which probably represents some thickness of soft beds).

9. On the high cliffs near Belfast are very bright red sandstones and shaly beds, with gray blotches and cylindrical fucoids—about....... 120

10. Over the last are seen, in the country east of Belfast, soft red sandstones with beds of conglomerate with rounded quartz pebbles and arenaceous cement (thickness uncertain)... 525

As seen in this section, the whole thickness of these beds cannot much exceed 500 feet. Of this the lowest 270 feet, being Nos. 1 to 5 inclusive, of the above section may be referred to the lower division, or " Bunter," and the remainder to the upper division of the formation, or " Keuper." The dips are so low, and the beds so much affected by oblique stratification, that those of the Trias cannot be said to be unconformable to the underlying Permian rocks; and for this reason, as well as on account of the similarity in mineral character between the two groups, some uncertainty may rest on the position of the line of separation. That above stated depends on fossils, or a somewhat abrupt change of mineral character, and on a slight change in the direction of the dip.

In the West, in the Rocky Mountains, and in British Columbia, the Trias is represented by argillites with bivalve shells of the genus *Monotis*, by red arenaceous rocks and also by large masses of diorite and felspathic igneous rocks. The gold veins of British Columbia are believed to be, in part at least, in the Triassic slates.

II. JURASSIC SYSTEM.

1. *Lias.* Not represented in Canada, unless some of the shales and sandstones overlying the Trias of Peace River are of this age. Not represented in the Eastern United States, unless some of the rocks referred to the Upper Trias are its equivalents. In England, the gray limestones and shales of Lyme Regis, &c., rich in Saurian remains. Similar beds with marls, &c., are extensively distributed in France, Switzerland, Italy and Suabia.

FOSSILS (in Europe)—This group is rich in marine shells. *Ammonitidæ* abound. *Pteroceras, Paludina* and other modern genera of gastropods appear. *Leptaena, Spirifera* and other old genera of Brachiopods become extinct. *Ostrea* and other recent forms of Lamellibranchs appear. *Enaliosaurs* are abundant and crocodiles of the genus *Teleosaurus* appear. Cycads and conifers are the most abundant fossil trees.

2. *Jurassic proper*, or *Oolitic Series.* Not represented in Canada, except perhaps by porphyrite and other volcanic rocks in British Columbia, and shales and sandstones of Rock Island, Peace River. Limestone and marl of Black Hills and elsewhere in Western United States. In Europe—Lower, Middle and Upper Oolite of England, with

the intervening Oxford and Kimmeridge clays. Also very largely represented in France and the Jura Mountains and elsewhere in Europe. The Lower or Bath Oolite of England is remarkable for oolitic structure. The Stonesfield slate, a flaggy series connected with the Lower Oolite, is noted for vegetable remains and remains of mammals and insects. The Lithographic slate of Solenhofen has many interesting fossils and is of the age of the Middle Oolite. It has afforded the earliest known bird, *Archæopteryx macrurus.* The Upper or Portland Oolite is overlaid by a fresh-water formation, the Purbeck, which has afforded many mammalian remains and land plants, and also fresh-water snails allied to *Planorbis.*

FOSSILS.—Remarkably rich in Cephalopods, especially *Ammonitidæ* and *Belemnitidæ.* Also in Reptiles, as *Pterodactyls, Dinosaurs, Enaliosaurs, Crocodileans, Turtles*, &c.

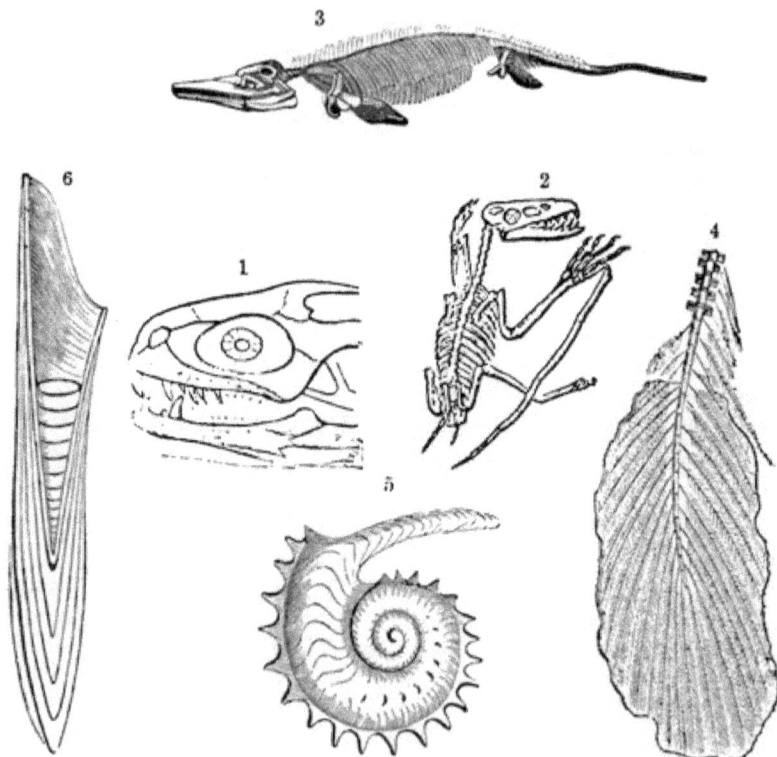

Fig. 143.—JURASSIC FOSSILS.

1.—Head of *Megalosaurus.* 2. *Pterodactylus crassirostris.* 3. *Ichthyosaurus communis.* 4. Tail of *Archæopteryx.* 5. *Ammonites Jason.* 6. *Belemnites* (section).

III. CRETACEOUS SYSTEM.

1. *Lower Cretaceous* or *Neocomian.* In Canada—Tatlayoco lake sandstone and conglomerate, with *Aucella Piochii*, and underlying porphyrites; and perhaps the coal series of Queen Charlotte Island and Quatseno Sound, in British Columbia; Shasta group in California; Lower Cretaceous clays of New Jersey, &c. In Europe—Hastings sand, Weald clay, and lower greensand of England, and their equivalents on the continent.

FOSSILS.—Dinosaurian Reptiles, *Iguanodon* and *Hylaeosaurus*, &c. Appearance of Teleost fishes, and of Angiospermous Exogens of modern types. *Crioceras, Ancyloceros,* and *Ammonites* abundant; *Diceras.*

2. *Middle Cretaceous* or *Cenomanian.* In Canada—Dakota group of Western Territories and its extension north of the 49th parallel; Niobrara limestones and clays of Western Territories and Western States of the Union. This is an extensive marine formation, rich in Foraminifera with *Ostrea congesta* and species of *Inoceramus* and *Baculites.* In Europe— the Gault clay, Upper Greensand and Chalk marl of England and the continent of Europe.

FOSSILS.—Species of *Hamites, Scaphites, Turrilites, Lima, Ostrea,* &c., are characteristic in Europe. Plants of modern types.

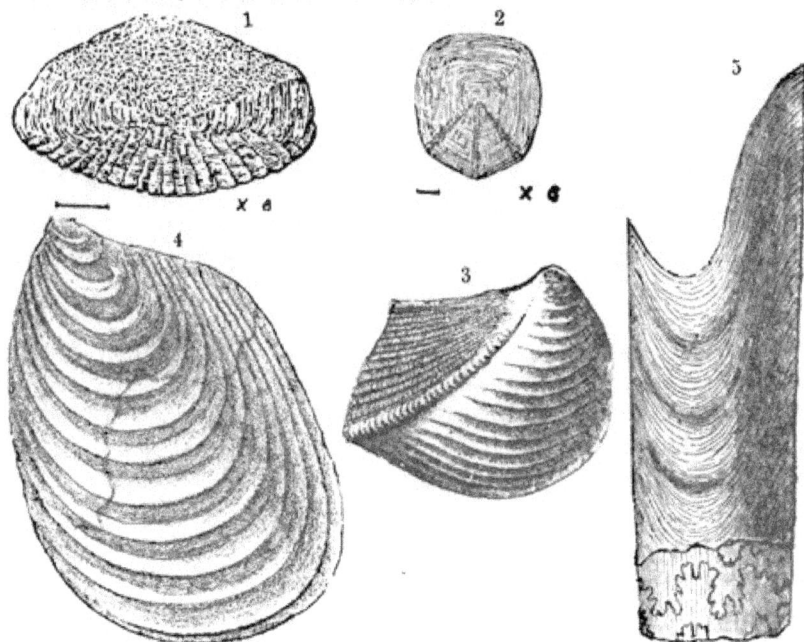

Fig. 144.—CRETACEOUS FOSSILS. (Western America.)
1 and 2.—Scales of Teleost Fishes, N. W. Territory. 3. *Trigonia Americana.*
4. *Inoceramus Vancouverensis.* 5. *Baculites ovatus.*

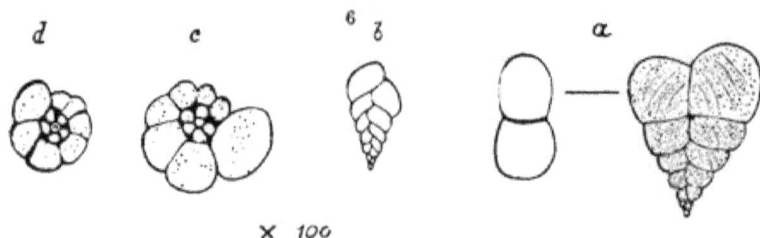

× 100

6. CRETACEOUS FORAMINIFERA, Boyne R., Manitoba,

(a) Textularia globulosa.   (b) T. pygmœa.   (c) Planorbulina ariminensis.
(d) Planorbulina.

3. *Upper Cretaceous* or *Senonian.*   In Canada—Ft. Pierre and Fox Hill clays and sandstones of the Western Territories, and continuation to the South.   Greensand of New Jersey with associated clay and limestone.   White or Upper chalk of England and other parts of Europe, and white limestones of North Africa and Western Asia, Mæstricht limestone of Denmark.

FOSSILS.—Vast numbers of Oceanic Foraminifera, especially *Globigerina ;* Coccoliths ; Sponges of genus *Ventriculites,* &c. ; Echinoderms of genera *Ananchites, Galerites, Marsupites, Cidaris,* &c. ; Lamellibranchs of genera *Inoceramus, Spondylus, Ostrea,* &c. ; Cephalopods of genera *Belemnitella, Baculites, Nautilus,* &c. ; Reptiles of genera *Mosasaurus, Clidastes, Hadrosaurus ;* toothed birds of genera *Icthyornis, Hesperornis,* &c.   The flora of this period contains a large preponderance of modern types.   Mammalian Remains (March 1889).

Fig. 145.—Jaw of a Cretaceous toothed bird *(Ichthyornis dispar.)*
After Marsh.   Natural size.

Distribution.—The Cretaceous rocks occupy a broad belt extending on the 49th parallel from near the Red River to the Souris River and thence to the north-west.   They also occur on the Saskatchewan and head waters of the Missouri farther to the west.   Considerable areas occur in British Columbia, the most important being that which includes the Nanaimo and Comox coal-field on Vancouver Island.

In the North-west territories of Canada a great breadth is occupied by cretaceous rocks which extend southward to the Gulf of Mexico, indicating an extensive inland sea East of the Rocky Mountains in the Cretaceous age.   The following section is that given by Meek and Hayden for these beds as they occur south of the U. S. boundary.   (order descending):—

No. 5. *Fox Hill Beds.*—Grey, ferruginous and yellowish sandstones and arenaceous
    clays. *Marine Shells.* 500

No. 4. *Fort Pierre Group.*—Dark-grey and bluish plastic clays. *Marine shells,*
    gypsum and fish remains. 700

No. 3. *Niobrara Group.*—Calcareous marls. *Marine shells, fish remains, foraman-*
    ifera, &c. 200

No. 2. *Fort Benton Group.*—Dark-grey laminated clays, with some limestone.
    Marine shells. 800

No. 1. *Dakota Group.*—Yellowish, reddish and whitish sandstones and clay, with
    occasional lignites. *Marine and some fresh-water shells and angiosperm-*
    ous leaves 400

The Cretaceous Period is remarkable, in both the eastern and western
continents, for a prevalence of estuarine and fresh-water conditions in its
earlier portion, and for a great subsidence, producing oceanic conditions
over wide areas now land, in its middle and latter portion. It is also
marked by the decadence of the reign of reptiles, and by the introduction
of the modern flora in the continents of the Northern Hemisphere.

North of the Canadian boundary, and especially westward, the Niobrara
and Benton groups of the above section change their character, and are
composed of sandy shales and sandstones, with beds of coal and fossil
plants, and shells indicating brackish water. This has been called * the
Belly-River Series, and contains important beds of lignite and coal. In
the prairies these deposits are nearly horizontal; but in the Rocky
Mountains they have been included in the folds of the older rocks so as
to form narrow troughs with high dips. Coincidently with these disturbances the coal is changed into anthracite.

On the west coast, in Vancouver Island, the cretaceous rocks constitute
the coal-bearing horizon of British Columbia. They have been folded
and disturbed in the movements in progress in Tertiary times, and have
no doubt been much reduced by denudation.

"Their most important area, including the coal-mining regions of
Naniamo and Comox, may be described as forming a narrow trough along
the north-east border of Vancouver Island, 130 miles in length. The
rocks are sandstones, conglomerates and shales. They hold abundance of
fossil plants and marine shells in some places, and in appearance and
degree of induration much resemble the true Carboniferous rocks of some
parts of Eastern America. In the Naniamo area the formation has been
divided by Mr. J. Richardson as follows, in descending order:—

        Sandstones, conglomerates and shales... 3290 feet.
        Shales... .... 660 "
        Productive Coal-measures.............. 1316 "
                             ————
                             5266 "

* Dr. G. M. Dawson, Reports Geol. Survey of Canada. Possibly the Lower
Laramie may also be Upper Cretaceous.—See next heading.

"In the Queen Charlotte Islands, Cretaceous rocks cover a considerable area on the east coast, near Cumshewa and Skidegate Inlets. At Skidegate they hold true anthracite coal, which, besides being a circumstance of considerable geological interest, would become, if a really workable bed could be proved, a matter of great economic importance to the Pacific coast.

At Skidegate, where these rocks are most typically developed, they admit of subdivision as follows, the order being, as before, descending:—

A.	Upper shales and sandstones	1500 feet.
B.	Coarse conglomerates	2000 "
C.	Lower shales with coal and clay ironstone....	5000 "
D.	Agglomerates	3500 "
E.	Lower sandstones	1000 "
		13,000 "

The animal fossils of the lower part of the Nanaimo beds have been determined by Whiteaves as equivalent to those of the Chico group in California, or to the Cenomanian and Senonian of Europe. In the Queen Charlotte Islands and on the mainland are beds containing *Aucella Piochii*, and regarded as equivalent to the Shasta group of California, or Upper Neocomian. In the Rocky Mountains are still older beds, the Kootanie series, holding fossil plants. These would probably occur below the Dakota group of the section given above.

The fossil plants of the Neocomian, as represented by the Kootanie series and the Queen Charlotte Island coal series, have a strictly mesozoic aspect, and consist of Conifers, Cycads and Ferns. But in the Dakota group and equivalent beds in Canada, we find numerous genera and species of Exogenous trees, as *Populus, Sassafras, Platanus, Aralia* and *Betula*. In the Upper Cretaceous of Vancouver Island there is a very rich angiospermous flora, with Conifers of the genus *Sequoia* and a fan palm *(Sabal)*. Thus the modern types of plants appear in the Middle and Upper Cretaceous, and then appear also the ordinary or teleost fishes, which in this age take the place of the ganoids, and present generic forms identical with those of the modern waters.

The useful minerals of the Cretaceous are the coals already referred to, and which are of the greatest economic importance in Vancouver Island, in the Rocky Mountains and in the Bow and Belly River Districts.

KAINOZOIC PERIOD.

I. Eocene Age.

1. *Lower Eocene (Orthrocene).* In Canada this formation is probably represented by the Lignite Tertiary formation of the Western Territories, the Lignitic or Laramie group of the American geologists, which consists of estuarine and fresh-water sandstones and shales, with reptilian remains, lignite and fossil plants of modern types. It is, however, regarded by many geologists as more nearly related to the Upper Cretaceous than to the Eocene proper. Fig. 141, and the section on p. 34, represent parts of

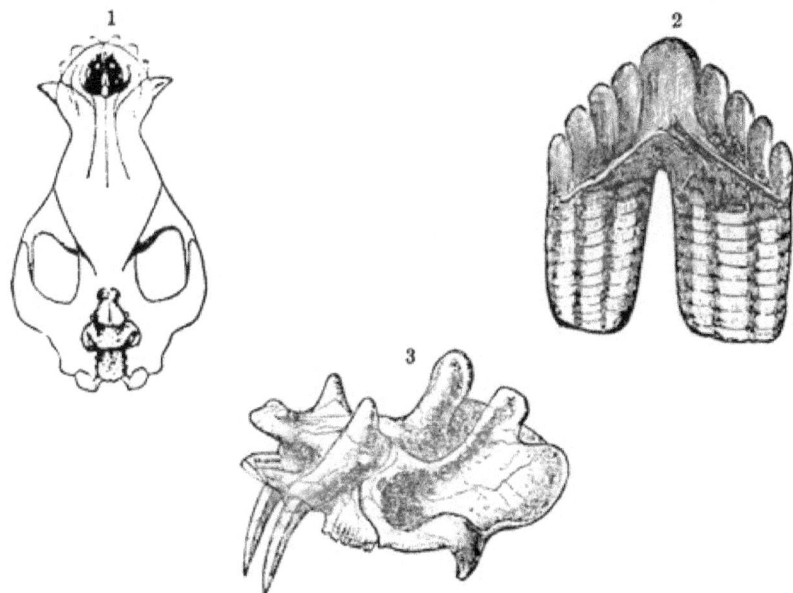

Fig. 146.—Kainozoic Mammals.

1.—*Coryphodon hamatus* (Eocene) reduced. 2. *Zeuglodon cetioides*—tooth (Eocene) reduced. 3. *Dinoceras mirabilis* (Eocene) reduced.

this group, which is very extensively distributed in the region between the Red River and the Rocky Mountains and thence southward. In England the typical formations are the London clay and Thanet sands, holding marine and estuarine shells and fossil fruits and wood. The Argile Plastique and Sable Bracheux represent it in the Paris basin. In these beds, in Europe, the oldest known placental mammals occur. (*Hyracotherium, Lophiodon, Coryphodon*). Marine invertebrates of living species also appear, though as yet in small numbers, about three per cent. of the whole.

2. *Middle Eocene,* or *Eocene proper (Nummulitic).*   Not as yet recognized in Canada.   In the United States this series is represented by the clays, marls, sands and coarse limestones of the Claiborne series of Alabama, holding marine shells and bones of *Zeuglodon.*   In the west the great lake basins of the Wahstach have afforded remains of many land animals *(Coryphodon, Tillodontia, Eohippus,* &c).   In Europe the most characteristic and wide-spread formations are the Calcaire Grossier of the Paris Basin and its associated marine sands, and the Nummulitic limestones extending from Western Europe to India, and marking a great subsidence.   In England the Bracklesham and Alum Bay series are of this age.

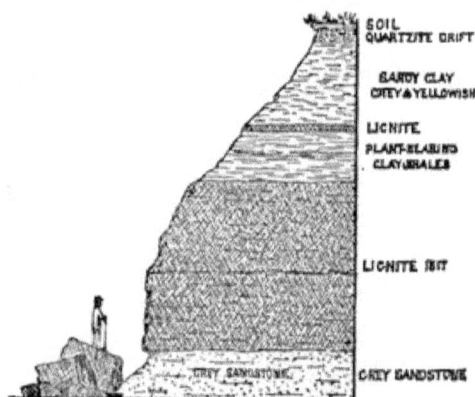

Fig. 147.—Lignite Bed.   Porcupine Creek, N. W. T.—(G. M. Dawson).

3. *Upper Eocene (Proicene).*   Not recognized in Canada.   In the United States represented by the marine clays and Orbitoidal limestone of Alabama, Mississippi, &c. (Vicksburg group), and in the west by freshwater clays and sands (Bridger group, &c.), containing *Dinoceras, Uintatherium, Orohippus,* &c.   In Europe the best known representative is the Gypseous series and Silicious Limestone of the Paris Basin, and the upper beds of the Isle of Wight series in England.   These formations abound in mammalian remains *(Anchitherium, Palæotherium, Anoplotherium, Xyphodon, Lemuridæ).*

A very equable and warm climate seems to have prevailed in the Eocene and Miocene, so that plants of genera now living in temperate climates were abundant in Greenland and Spitzbergen.

The Laramie of the North-West Territories rests conformably on the Upper Cretaceous and has thick beds of lignite and many fossil plants, with fresh-water shells of the genera *Unio*, *Goniobasis* and *Viripara*. Its general distribution and structure may be stated as follows :—

On the geological map of Canada, the Laramie series, formerly known as the Lignitic or Lignite Tertiary, occurs, with the exception of a few outliers, in two large areas west of the 100th meridian, and separated from each other by a tract of older Cretaceous rocks, over which the Laramie beds may have extended, before the later denudation of the region.

The most eastern of these areas, that of the Souris River and Wood Mountain, extends for some distance along the United States boundary, between the 102nd and 109th meridians, and reaches northward to about thirty miles south of the "elbow" of the South Saskatchewan River, which is on the parallel of 51° north. In this area, the lowest beds of the Laramie are seen to rest on those of the Fox Hill group of the Upper Cretaceous, and at one point on the west they are overlaid by beds of Miocene Tertiary age, observed by Mr. McConnell, of the Geological Survey, in the Cypress Hills, and referred by Cope, on the evidence of mammalian remains, to the White River division of the United States geologists, which is regarded by them as Lower Miocene.* The age of the Laramie beds is thus stratigraphically determined to be between the Fox Hill Cretaceous and the Lower Miocene. They are also undoubtedly continuous with the Fort Union group of the United States geologists on the other side of the international boundary, and they contain similar fossil plants. They are divisible into two groups—a lower, mostly argillaceous, and to which the name of "Bad Lands beds" may be given from the "bad lands" of Wood Mountain where they are well exposed, and an upper, partly arenaceous member, which may be named the Souris River or Porcupine Creek division. In the lower division are found reptilian remains of Upper Cretaceous type, with some fish remains more nearly akin to those of the Eocene.† Neither division has as yet afforded mammalian remains.

The western area is of still larger dimensions, and extends along the eastern base of the Rocky Mountains from the United States boundary to about the 55th parallel of latitude, and stretches eastward to the 111th Meridian. In this area and more especially in its southern part, the officers of the Geological Survey of Canada have recognized three divisions

---

* Report of Geological Survey of Canada, 1885.

† Cope in Dr. Dawson's Report on 49th parallel.

as follows :—(1) The Lower Laramie or St. Mary River series, corresponding in its character and fossils to the Lower or Bad Lands division of the
other area. (2) A Middle division, the Willow Creek beds, consisting of
clays, mostly reddish, and not recognized in the other area. (3) The
Upper Laramie or Porcupine Hills division, corresponding in fossils and
to some extent in mineral character to the Souris River beds of the
eastern area.

The most characteristic plant remains of the Upper and Lower
Laramie respectively are the following :—*Platanus nobilis, P. Raynoldsii,*
&c.; *Corylus McQuarrii, Populus Arctica,* and other species, and species of
Salix, Ulmus, Sassafras, Viburnum, &c. ; also *Thuja interrupta, Sequoia
Nordenskioldi, Taxites Olriki,* &c., and the ferns *Onoclea sensibilis* and
*Davallia tenuifolia.**

Very thick and valuable beds of Lignite occur in the Laramie series of
the North-western Territories.

II. MIOCENE AGE.

1. *Lower Miocene* or *Oligocene.* The Fort Union Group, or Upper
Laramie, near the Canadian boundary in the U. States, once regarded as
Miocene, is now known to be older, and indeed many groups of beds
holding plants of Eocene age in Europe and North America, as well as in
Greenland, have been wrongly referred to the Miocene.

So far as known this group is very slenderly represented in Canada.
A small area occurs in the Cypress Hills in the Territory of Assinaboia.
It consists of conglomerate, soft sandstone and clay; and has afforded
some mammalian remains, considered as belonging to the White River
series or Lower Miocene of the United States. The Miocene may also
be represented by the volcanic rocks, sandstones and shales with lignite
and fossil plants of Nicola, Similkameen, &c., in the interior of British
Columbia. In America the subdivisions of the Miocene have not been
distinctly separated, but the age is represented by the New Jersey,
Virginia, &c., middle Tertiary sands, clays, marls and infusorial deposits ;
and in the West by the Middle Tertiary lake basins of White River, &c.,
east of the Rocky Mountains. In the latter, three subdivisions are characterized by Marsh as respectively those of the *Brontotherium, Oreodon*
and *Miohippus.* The Miocene beds hold a larger percentage of recent
shells than the Eocene (17 to 30 per cent.), and abound in mammalian

---

* See papers by the author in Trans. R. S. C., 1886.

remains *(Brontotherium, Titanotherium, Oreodon, Machairodus, &c.)* In England—Hempstead beds of I. of Wight, Bembridge and Headon beds. In France—Calcaire de Beauce and Sables de Fontainebleau, with equivalent deposits in Germany, Italy, &c. The basalts of Antrim and the Hebrides are, in part, of this age. Living genera of mammals, as *Rhinoceros, Tapirus, Mustela, Sciurus*, &c., appear.

2. *Middle Miocene.* Falunien of France, Middle or marine Molasse sandstone of Switzerland. Genera *Mastodon, Dinotherium, Sus, Antilope, Cervus, Felis, Dryopithecus, &c.*

3. *Upper Miocene.* In Europe, Molasse of Oeningen in Switzerland, Léberon and Epplesheim beds of France, Pikermi formation in Greece. Additional modern genera of mammals, as *Camelopardalis, Gazella, Hyæna* and *Hystrix* appear.

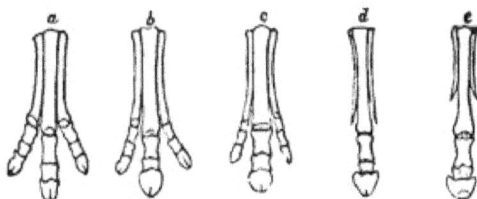

Fig. 148.—Series of Equine Feet.—After Marsh. *a, Orohippus*, Eocene. *b, Miohippus*, Miocene. *c, Protohippus*, Lower Pliocene. *d, Pliohippus*, Upper Pliocene. *e, Equus*, Post-Pliocene and Modern.

Fig. 149.—*Oreodon major* (Miocene) reduced.

III. PLIOCENE AGE.

1. *Older Pliocene.* Not recognized in Canada. In the United States, Sumter clays and sands of North and South Carolina. In the West, Loup River group of Niobrara, containing remains of Camel, Rhinoceros, Horse, &c. Sandstones of Los Angeles, California, and fresh-water basins of Oregon. The remarkable phosphatic deposits of South Carolina, which contain vast numbers of bones of animals of various ages swept together by water, have been referred to this age. In England, Coralline Crag and Red Crag.

2. *Newer Pliocene.* Not recognized in Canada, nor distinguished in the United States from the older Pliocene. In England, Norwich Crag and Chillesford clay.

In the Pliocene, the percentage of recent shells rises to 50 or more. Mammalia of modern genera are abundant, and a few modern species appear. In the later Pliocene the land both of Europe and America seems to have been more elevated and extensive than at present (First Continental period of Lyell). The climate of the Northern Hemisphere was cooler than in the Miocene.

The Pliocene, having been in the main a time of continental elevation, the deposits of this age are usually confined to the margins of the continents, or occur in inland lake basins. There is evidence that in this age the Northern Hemisphere was occupied by a very grand and varied mammalian fauna, including several species of *Mastodon*, *Elephas* and *Rhinoceros*, as well as animals allied to the Camel, Horse and other modern genera.

IV. Pleistocene Age.

This was characterized throughout the Northern Hemisphere by a great refrigeration of climate, followed or accompanied by a submergence of the land to a depth exceeding in some places 5000 feet. The formations of the period are well represented in Canada, and may be taken as types, more especially as from their great extent and uniformity they are free from some of the complications which have caused controversy elsewhere.

1. *Boulder Clay.* At the beginning of the Pleistocene the land was higher than at present. At this time the mountain tops were extensively occupied with glaciers, which have left their traces in all the elevated ground. Very deep valleys and ravines were also excavated by the rivers. Beds of peat were accumulated, and gravels and sands in low grounds, in lake basins and on coasts. Gradual subsidence then set in, under which the valleys were invaded by cold Arctic currents laden with field ice and bergs, while the high levels still sent down glaciers. Under these circumstances moraines were formed on the land, and sheets of stony clay with boulders in the sea, forming what has been termed the boulder clay or "Till," and extensively polishing and striating the surface of rocks.

In the deposits of this period Arctic shells are found, though not abundantly, and also trunks of boreal coniferous trees. At the beginning of the age, however, there were in Europe and America forests of temperate and boreal type, and a great number of mammals, some extinct, some still surviving, and presenting a remarkable mixture of boreal and temperate forms. Remains of these occur in peaty beds under the boulder clay. By the progress of the glacial cold and subsidence, these animals were destroyed or compelled to migrate to the southward.

Fig. 150.—Boulder (11 feet long) on glaciated surface. Lake of the Woods.

2. *Leda Clay, Erie Clay.* This marks the greatest subsidence and the gradual emergence of the land. It is a fine stratified clay, sometimes, however, with large boulders, and thus passing into boulder clay. It has on the Atlantic slopes of America and Europe numerous marine fossils, especially in its upper part ; and these are mostly of species still inhabiting the North Atlantic and North Pacific. Farther inland it contains some remains of plants and land animals. The Leda clay is equivalent to the Clyde beds and Uddevalla beds of Europe. There is reason to believe that the great subsidence which closed in this period reached to 2300 feet in the mountains of Wales, and to 4000 feet in those of North America. It was probably greatest toward the north. At the beginning of the deposit of Leda clay, the shells indicate cold water covered with floating ice. Toward its close (Upper Leda clay or Uddevalla beds) the marine climate must have been little different from that now occurring in the same latitudes on the western side of the Atlantic.

3. *Saxicava Sand and Second Boulder Drift.* This marks the re-elevation which ended in a second continental period, raising the continents to a greater elevation than at present The climate was still somewhat cold, and large boulders carried by floating ice abound in the Saxicava sand, but are most abundant at its base.

The following may be taken as a complete synopsis of the Pleistocene of Eastern Canada in ascending order :—

a. Peaty terrestrial surface anterior to boulder clay.

b. Lower stratified gravels—(Syrtensian deposits of Matthew).

c. Boulder clay and unstratified sands with boulders. Fauna, when present, extremely Arctic.

d. Lower Leda Clay, with a limited number of highly Arctic shells, such as are now found only in permanently ice-laden seas.

e. Upper Leda clay and sand, or Uddevalla beds, holding many sub-Arctic or boreal shells similar to those of the Labrador coast at present. This in some places contains remains of boreal plants and corresponds to what has been termed in the west "Interglacial."

f. Saxicava sand and gravel, either non-fossiliferous, or with a few littoral shells of boreal or Acadian types. This often contains travelled boulders and constitutes an upper or newer Boulder deposit.

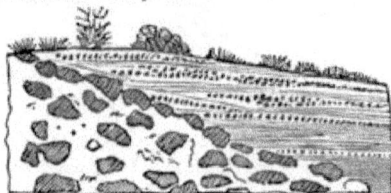

Fig. 151.—Stratified Gravel resting on Boulder Deposit.   Nova Scotia.

Fig. 152.—Modern Ice-drift.   Travelled Stone resting on recent tidal mud, Petitcodiac River.

The Pleistocene deposits are sometimes called *Quaternary ;* but there is no good ground for separating them from the Kainozoic or Tertiary. The term "Champlain" deposits has been applied to them in the United States ; but the Lake Champlain beds are those of a limited valley among mountains and are not typical or characteristic.

Some writers include in the Pleistocene the next or post-glacial age ; but it is more nearly connected in its physical conditions and its animal life with the modern period.

Dawkins catalogues, for this period in Britain, 1 mammal surviving from the Pliocene and still living, 7 surviving from Pliocene and extinct, 67 new species, of which 14, including elephants and other large and important species, are now extinct.

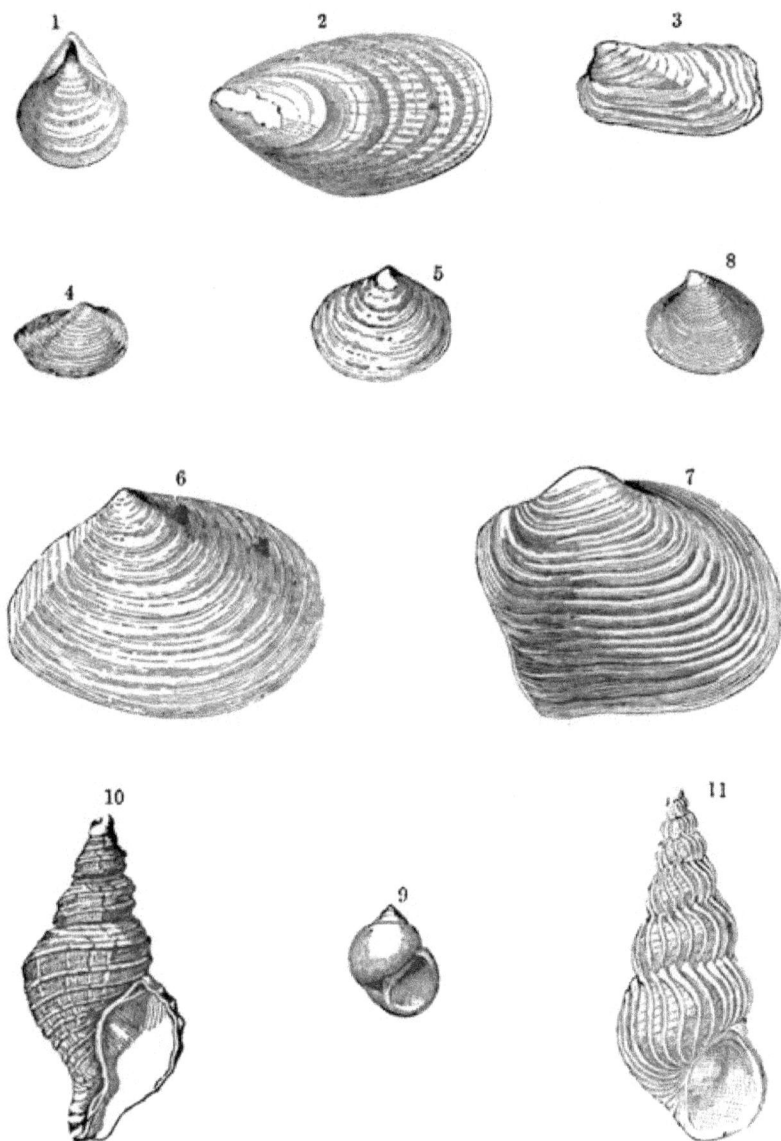

Fig. 153.—PLEISTOCENE FOSSILS.

Fig. 1.—*Rhynchonella psittacea.* 2. *Mytilus edulis.* 3. *Saxicava rugosa,* 4. *Leda (Portlandia) arctica.* 5. *Tellina (Macoma) Grœnlandica.* 6. *Tellina (Macoma) alcarea.* 7. *Mya truncata.* 8. *Astarte (Nicania) Laurentiana.* 9. *Natica clausa.* 10. *Fusus tornatus (Neptunea despecta).* 11. *Scalaria Grœnlandica.*

The following description of the drift of the western plains is from a paper by Dr. G. M. Dawson :—*

"Resting immediately on the surface of the cretaceous and Laramie rocks in a number of localities on the Bow, Belly, Old Man, and other rivers, is a deposit of well-rolled pebbles of shingle, consisting, for the most part, of hard quartzites, and derived entirely from the palæozoic rocks of the Rocky Mountains. These pebbles are seldom more than a few inches in diameter, and often very uniform in size. The deposit has been observed to extend to a distance of over a hundred miles from the base of the mountains. Whether it has been carried from the mountains entirely by the action of rapid streams of preglacial times, or has been distributed in some more extended body of water, I am as yet unprepared to decide ; but the fact that it occurs at very different elevations above the present water-level in neighboring sections on the same river, would appear to point to the latter conclusion. No marks of ice-action have been found on the stones of this deposit, which at one place on the Belly was observed to be associated with stratified sand beds.

Resting upon the shingle deposit in some localities, but in other places directly on the Cretaceous and Laramie, is the Boulder-clay, a mass of sandy clay, often very hard, and not infrequently showing a pretty well marked relation in colors and material to the underlying soft rocks, from which it has evidently been largely formed, but packed irregularly with boulders and fragments of Laurentian and Huronian origin, often distinctly glaciated, and with quartzite pebbles resembling those above described. While generally rather massive in character, the boulder-clay is frequently more or less evidently divided by stratification planes, and is quite distinct in appearance from the morainic accumulations which occur in the foot-hill belt.

The upper part of the boulder-clay is usually much more distinctly stratified than the lower, and often more or less markedly lighter in color, though still holding numerous stones and boulders of mingled Laurentian and Rocky Mountain origin. In the region through which the lower part of the Belly River cuts, a series of well-stratified sands and sandy clays are intercalated between these two divisions of the boulder-clay ; and in several sections these were observed to include an irregular layer of impure lignite or indurated peat a few inches in thickness, evidently the accumulation in a swamp or shallow lake which must have covered

---

* Quart. Journ. Geol. Soc., Nov., 1875. Geology and resources of the 49th parallel.

many miles of surface. A thin nodular deposit of ironstone was also found in association with the lignite at one place."

The Missouri Coteau is a great ridge of drift deposits at the edge of the third or Tertiary plateau of the prairie, about 400 miles west of the Laurentian area, from which it is separated by the palæozoic district of Manitoba and by the first and second prairie plateaus. It is thus described by Dr. G. M. Dawson :—

" The Missouri Coteau is one of the most important features of the western plains, and is certainly the most remarkable monument of the Glacial period now existing there. I have had the opportunity of examining more or less carefully that portion of it which crosses the forty-ninth parallel, north-westward for a length of about 100 miles. On the parallel, the breadth of the Coteau, measured at right angles to its course, is about 30 miles ; and it widens somewhat northward.

On approaching its base, which is always well defined at a distance, a gradual ascent is made, amounting in a distance of 25 miles to over 150 feet. The surface at the same time becomes more markedly undulating, as on nearing Turtle Mountain from the east, till, almost before one is aware of the change, the trail is winding among a confusion of abruptly rounded and tumultuous hills. They consist entirely of drift material ; and many of them seem to be formed almost altogether of boulders and gravel, the finer matter having been to a great extent washed down into the hollows and basin-like valleys without outlets with which this district abounds. The ridges and valleys have in general no very determined direction ; but a slight tendency to arrangement in north-and-south lines was observable in some places.

The boulders and gravel of the Coteau are chiefly of Laurentian origin, with, however, a good deal of the usual white limestone and a slight admixture of the quartzite drift. The whole of the Coteau-belt is characterized by the absence of drainage valleys ; and in consequence its pools and lakes are often charged with salts, of which sulphates of soda and magnesia are the most abundant. The saline lakes frequently dry up completely towards the end of the summer, and present wide expanses of white efflorescent crystals, which contrast in colour with the crimson *Salicornia* with which they are often fringed.

Taking the difference of level between the last Tertiary rocks seen near the eastern base of the Coteau, and those first found on its western side, a distance of about 70 miles, we find a rise of 600 feet. The slope of the surface of the underlying rocks is, therefore, assuming it to be uniform, a little less than 100 feet per mile.   On and against this gently inclined plane the immense drift deposits of the Coteau hills are piled.

The average elevation of the Coteau above the sea, near the 49th parallel, is about 2000 feet ; and few of the hills rise more than 100 feet above the general level."

The Missouri Coteau can be traced across the central region of North America for 800 miles, and is believed by the geologists of the United States to be traceable south of the great lakes to the Atlantic coast, and to constitute what they regard as the moraine of a great confluent glacier. In the North-west however, where it attains to its maximum development, it is evidently the edge of a shoal or coast, on which floating ice has for a long period discharged its burden of boulders and debris.

V. Modern Age.

This extends from the close of the glacial or Pleistocene age to the present time, and is divisible into two well-marked periods.

1. *The Post-Glacial.*   (Second Continental Period of Lyell.)   In this the land of the Northern Hemisphere was more extensive than at present. The climate was temperate but somewhat extreme.   All the modern mammals, including man, seem to have been in existence, but several others now extinct, as the Mammoth, the Tichorhine Rhinoceros and the Cave Bear, lived in the Northern Hemisphere, and many still extant differed very remarkably in their geographical distribution from that of the present time.   To this period belong the human remains of the early cave deposits and river gravels of Europe, or of the "Mammoth age" (Palæocosmic or Palæolithic age.)   This period was terminated by a submergence or a series of submergences which with their accompanying physical changes proved fatal to many species of animals and to the oldest races of men, and left the continents at a lower level than at present from which they have risen in the recent period.   In Britain, Dawkins catalogues 22 living and 6 extinct species survivors of the Pleistocene in this period, and 18 new forms still living.   The 6 extinct species include 2 species of elephant, 2 of rhinoceros, the cave bear and the great Irish elk.   It is evident therefore that man comes in with a fauna in the main modern, but including a few large and important species which have perished since his advent, and many others which have much changed their range.

2. *The Recent* or *Historic Period.* This dates from the settlement of our continents at the present levels after the Post-glacial subsidence. It is the period of Neocosmic or Neolithic men of races still extant. I have called this the Historic Period, because in some regions history and

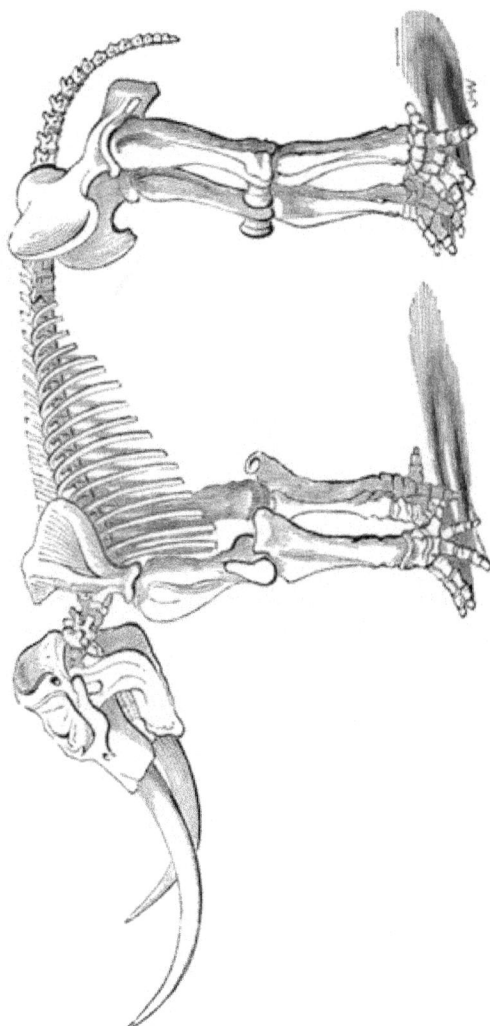

Fig. 154.—*Mastodon Americanus,* (Pleistocene and Modern.)

tradition extend back to its beginning. The historical deluge is in all likelihood identical with the movements of the land above referred to, by which this age was inaugurated; though in certain localities, as in

America, the beginning of the historic period is very recent. In this age man coexists wholly with existing species of mammals, and the races of men are the same which still survive. The whole forms geologically one period, and the distinctions made by antiquarians between stone, bronze and iron ages, and under the former between palæolithic and neolithic, are merely of local significance, and connected with no physical or vital changes of geological importance. The real geological distinction is that of Palæocosmic, Post-glacial or Antediluvian man on the one hand and Neocosmic, Recent or Post-diluvian man on the other. The Palæocosmic men have been divided into two races, the Canstadt or Neanderthal type and the Engis or Cro-magnon type. Both of these were contemporaneous with the Mammoth, the tichorhine Rhinoceros, and other Post-glacial animals now extinct. It is probable that they may be ultimately identified with the ruder tribes of the historical antediluvian period, and that the physical changes by which they and some other animals seem to have been destroyed were the same with those recorded in the ancient history and tradition of all the older races of men.

Slight changes of level have occurred in the continents in the modern period. In Canada the most interesting evidence of these exists in the remains of submerged forests in the Bay of Fundy.* (Fig. 155.)

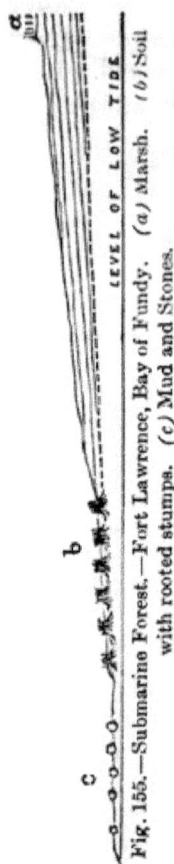

Fig. 155.—Submarine Forest.—Fort Lawrence, Bay of Fundy. (a) Marsh. (b) Soil with rooted stumps. (c) Mud and Stones. LEVEL OF LOW TIDE

* See Acadian Geology. p. 38

PART III. CANADIAN TOPOGRAPHY AND GEOLOGY.

The Dominion of Canada consists of a belt of more or less inhabited country extending for about three thousand miles across the continent of North America, between the parallels of 42° and 55° North Latitude, together with a vast region to the north of this belt not colonized and extending to the shores of the Arctic sea. Its most eastern part is Scatari Island, in W. Longitude 60° 25′, and its most western part, Graham Island, in the Queen Charlotte group, extending to W. Longitude 133°. Its total area is estimated at 3,406,562 square miles. Its area is thus rather less than that of Europe and somewhat greater than that of the United States, and it embraces representatives of nearly all the geological systems from the Laurentian to the Modern.

For the purposes of Geological description, Canada and Newfoundland may be divided into six regions, as follows :—

1. *The Acadian Region*, comprehending the Atlantic Provinces of Nova Scotia, New Brunswick and Prince Edward Island. These constitute a part of the Atlantic Slope of North America.

2. *The Canadian Region proper, or that of Quebec and Ontario*, including the two Provinces above named. These occupy the St. Lawrence and Ottawa valleys and their vicinity, and a portion of the basin of the great lakes, and belong to the Eastern and Northern part of the continental plateau of North America.

3. *The Manitoban and Northwest Region* may include the Province of Manitoba and the great plains extending westward to the Rocky Mountains, and including the territories of Assiniboia, Saskatchewan, Alberta and Athabasca.

4. *The British Columbian Region*, being the Province of British Columbia, extending across the Rocky Mountains and Cordillera ranges, with the Pacific coast and the islands adjacent.

5. *The Hudsonian or Arctic Region*, including all the vast territories stretching northward to the Arctic sea, and heretofore constituting the territory of the Hudson's Bay Company. This constitutes a portion of the Arctic Geological basin.

6. *The Terranovan Region*, or the great island of Newfoundland, which is geologically related on the one hand to the Acadian region and on the other to the Canadian.

It will be convenient to consider these several regions separately.

I. THE ACADIAN REGION.*

The general orographic features of this region consist of (1) the broken hilly ranges of Nova Scotia, extending along the Atlantic coast of that Province and into the Island of Cape Breton ; (2) the isolated ridge of the Cobequid hills, extending from west to east and joining the former eastwardly ; (3) the hilly range of Southern New Brunswick, stretching along the coast and sending off a wide branch to the North-east ; (4) the high ranges extending along the south side of the St. Lawrence, and separating the Acadian Region from the Canadian proper. This constitutes a northern extension of the Apalachian mountains, sinking however in Gaspé below the waters of the Gulf of St. Lawrence. The highlands of Nova Scotia and New Brunswick do not attain a greater elevation than about 1600 feet, but the northern ridge attains a much greater height beyond the limits of New Brunswick. Between these hilly districts are included the Silurian and Devonian area of Northern New Brunswick, the large triangular Carboniferous area in the centre of that Province, the Carboniferous and Triassic districts of Nova Scotia, and the depression occupied by the Bay of Fundy and its branches.

The first and fourth of the ranges above mentioned extend farther to the North-east than the others, and include between their extremities the semi-circular Acadian Bay of the Gulf of St. Lawrence in which lies the low and crescent-shaped Island of Prince Edward.

The whole of this Acadian region is characterized, like other parts of the Atlantic slope of North America, as distinguished from its interior plains, by a varied and uneven surface, and by great variety of soil and

---

* As Canadian Geology was referred to for illustrative examples in Part II. the statements made there and figures, etc. of fossils given will not be repeated in Part III., except when absolutely necessary for the connection. For the Acadian Region additional details will be found in the latest edition of the author's "Acadian Geology," and in recent Reports of the Geological Survey.

mineral products. In the latter, the Acadian Provinces are especially rich ; and in these and their maritime situation, they bear to the inland regions of Canada much the same relation with that which the British Islands bear to the plains of Central Europe. Nova Scotia, more particularly, is most richly endowed with coal, iron, and gold.

The formations represented in the Acadian Region are those from the Laurentian to the Trias, inclusive, with the addition of the Pleistocene deposits. The development of the older rocks corresponds very closely with that on the opposite side of the Atlantic, and differs materially from that in the interior plateau of America. We may therefore notice the several systems of formations in detail, and shall be enabled to point out their resemblances and differences as compared with those of the Canadian region, proper, next to be considered.

1. *The Laurentian.*—These rocks, as they occur near St. John, New Brunswick, have been arranged by Messrs. Bailey and Matthew, in their recent Reports, in a Lower and Upper series.[*]

The former consists, in ascending order, of gray, red and gray, and dark-gray gneiss, with chloritic gneiss and diorite. The latter consists of lime-stone, with graphite and serpentine, gray quartzites and diorite, gray slates and limestones with diorite. In one of the Upper Limestones I have recognized somewhat obscure structures, which appear to indicate the presence of fragments of Eozoon.[†]

In Cape Breton the gneisses of St. Anne's Mountain resemble the Lower Laurentian of Canada, and the evidence that they may be of this age has been much strengthened by the recent observations of Mr. Fletcher. Specimens, and the observations of Mr. Brown and Mr. Campbell and others, induce me also to believe that in the little island of St. Paul, and in some parts of Northern Cape Breton, we may have a continuation of the rocks referred by Mr. Murray to the Laurentian in Newfoundland. With these exceptions, I have not seen in Nova Scotia, unless in travelled boulders, any rock that I could believe to be litho-logically equivalent to the Laurentian of Canada, nor have I found any stratigraphical evidence of the occurrence of such rocks.

2. *The Huronian.*—The Coldbrook series of Messrs Bailey and Matthew, rising from beneath the Cambrian fossiliferous slates, has been referred to this age.

---

[*] Geol. Reports, 1871, etc.
[†] Proceedings American Association, 1876.

It appears from Reports of Bailey and Matthew that the Coldbrook or
Huronian series rests unconformably on the Laurentian, and that pebbles
of the latter are included in its conglomerates.   On the other hand, the
Acadian or Cambrian beds lie unconformably on the Coldbrook series.
There would appear to be indications of unconformity between the upper
and lower members of the Huronian itself.   It thus appears as a distinct
formation, or group of formations, between the Laurentian and Acadian
groups, and connected with neither.   The study of a series of typical
specimens, kindly furnished by Mr. Matthew of St. John and Mr.
Murray, director of the Geological Survey of Newfoundland, enables me
to affirm a remarkable similarity in mineral character between the rocks
described as Huronian in these two regions, while their relations to the
Laurentian below and Cambrian above are the same.   The peculiar fossils,
however, *Aspidella* and *Arenicolites*, allied to *A. spiralis*, discovered by
Murray in the Upper Huronian of Newfoundland and described by
Billings, have not yet been recognized in New Brunswick.

The Huronian system, while consisting largely of conglomerates, hard
slates and quartzites, is remarkable for the abundance in it of felsites,
felsitic breccias, porphyry, diorite, and other crystalline or cryptocrystalline rocks, which, though stratified, are evidently of volcanic origin, and
if such rocks are to be considered as everywhere of this age, the classification of the older rocks of Acadia would be greatly simplified.   It is
evident, however, that we must separate some of these as probably of
later age.   There seems good reason to class as Huronian, or at least as
Lower Cambrian, the rocks of the Boisdale Hills in Cape Breton, which
Mr. Fletcher finds to underlie the fossiliferous Cambrian of that region,
and which are more quartzose and micaceous than the rocks of the
Cobequid series.   It is not impossible that rocks of this age may also
occur in the vicinity of the Cambrian beds found at Miré.   We may also
conjecturally class as Huronian the chloritic rocks of Yarmouth, N.S.

3, *The Cambrian.*—Under this head we may place the interesting
" Acadian series " of St. John, so well characterized by its fauna, that it
may be considered as the typical representative in Eastern America of
that Middle or Lower Cambrian formation known in England as the
Menevian, and of Barrande's étage C of the Primordial in Bohemia.

Matthew has recently greatly extended our knowledge of these beds
and has shewn that they include several subdivisions of the Lower
Cambrian.   These subdivisions are given by him in a recent paper and
may, with slight modification, be arranged as follows, in ascending order :

1. *Lower Cambrian.*—Basal or Etcheminian series of St. John, Lower Cambrian of Cape Breton and Newfoundland ; Georgian series, Vermont ; Caerfai of Wales, and Fucoid and Eophyton sandstones in Sweden.

FOSSILS.—*Obolus pulcher*, Worm burrows, *Ellipsocephalus*, *Olenellus*, &c. *

2. *Middle Cambrian.*—Acadian or St. John series of New Brunswick, Paradoxides beds of Newfoundland, Massachusetts and Sweden ; Solva, Menevian and probably Lower Lingula Flag beds of Wales.

FOSSILS.—*Paradoxides*, *Conocoryphe*, *Ptychoparia*, and many other Trilobites, *Hyolithus*, *Stenotheca*, *Orthis*, *Lingulella*, *Eocystites*, &c. This division holds the richest Cambrian fauna.

3. *Upper Cambrian.*—Not yet recognized in New Brunswick, but may be represented by some beds overlying the last. Potsdam sandstone and Calciferous of Quebec, Upper Lingula? and Tremadoc of Wales ; *Ceratopyge* and *Dictyonema* beds in Sweden.

FOSSILS.—*Dikellocephalus*, *Ceratapyge*, *Scolithus*, *Dictyonema sociale*, &c.

In the St. Lawrence valley the Upper Calciferous constitutes a transition group to the Siluro-Cambrian.—(See the description of that region.)

The great Atlantic coast series of Nova Scotia, which is the auriferous formation of that province, and includes, in ascending order, the so-called Quartzite and Clay-slate formations, in which these rocks respectively predominate, I believe to be likewise Cambrian or Primordial, a view which Dr. Selwyn and Professor Hind have also advocated. It is probably an equivalent of the Basal Cambrian of New Brunswick and of the older slate series of Newfoundland.

The evidence of fossils in determining the precise age of these rocks is unfortunately as yet somewhat imperfect. Dr. Selwyn has recognised in the Slate formation at Lunenburg linear markings of the nature of those which in Sweden have been named *Eophyton*† and have been described as land plants. They are, however, of very doubtful origin, and in my judgment more akin to those trails of aquatic animals which I have named *Rhabdichnites.* These beds also contain rounded bodies with radiating branches known as *Astropolithon.* (Fig. 156.)

---

* Matthew includes the Paradoxides beds in the lower member.

† Report of Geological Survey, 1870.

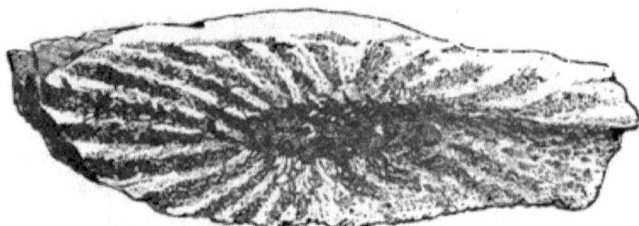

Fig. 156.—*Astropolithon Hindi.* Waverly, Nova Scotia.

The only other fossils known to me are specimens resembling the tubes or perforations named *Scolithus*, and very characteristic of the Potsdam sandstone in Canada. These I found near the mouth of St. Mary's River, in loose blocks, which must have been derived from a neighbouring ledge of quartzite. In so far as the above fossils give any indication of age, they serve to confirm the supposition that the Gold series of the Atlantic coast is to be referred to the Lower Cambrian.

Professor Hind has given a good description of the characters and structure of the more important parts of the Gold series in his Reports to the Department of Mines in Nova Scotia.† He states the entire

Fig. 157.—Vein of Auriferous Quartz, *a*, conformable to bedding. Waverly.

thickness of the series at 12000 feet. Of this the Lower or Quartzite and Slate division, which includes in its middle and upper part the productive gold veins, comprises about 9000 feet, and the Upper or Ferruginous Slate division 3000 feet. The whole is thrown into a series of somewhat sharp anticlinals, which, as might be expected, are much faulted. The steepest sides of the anticlinals are usually to the north,

† Report on Waverly District, 1869 ; Sherbrooke, 1870 ; Mount Uniacke, Oldham and Renfrew, 1872. (See also later papers by Mr. Poole and Mr. Gilpin.)

though in some cases, as at Sherbrooke, to the south. The courses of these anticlinals are approximately east and west. The gold has been found to be most accessible in the sides and near the summits of the anticlinals, while in the synclinals the upper unproductive slates usually appear. It is also to be observed that the productive gold veins are best developed in the vicinity of the great masses of eruptive granite of Devonian age which traverse this formation, and in connection with which it has locally been much metamorphosed, the slates assuming the character of Mica Schists with Staurolite and Chiastolite.

The gold veins, as stated in "Acadian Geology," run for the most part parallel to the bedding, but cross courses and branches traversing the beds are very frequent, and there is no proof that these are less ancient than the conformable veins or "leads." Though occurring in the Quartzite division, the auriferous veins usually follow bands of slaty rock included in the quartzite, a circumstance which much favours their profitable working.

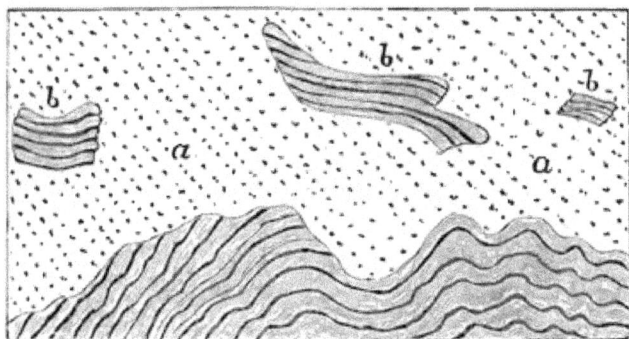

Fig. 158.—Junction of Granite and Slate, Nictau.

Intrusive granites occupy a considerable space in the Cambrian area. A map kindly communicated to me by Mr. Gilpin, F.G.S., shows that the largest granite mass extends continuously and with great breadth, from Shelburne northward into Annapolis county, and there bends eastward, and southward till it reaches the Atlantic, west of Halifax. Smaller isolated areas occur to the eastward of Halifax and in Guysboro county. All these granites are of Devonian age, and they alter the slates in their vicinity into Mica Schists, and cause the development of Staurolite, Chiastolite and Garnets; while the impure quartzites are converted into

I

imperfect gneisses. These changes are more manifest where numerous veins of binary granite penetrate the beds than where the latter are merely in contact with massive granite.

Mr. Fletcher, of the Geological Survey, has discovered, in certain beds near St. Andrew's Channel, Cape Breton, fossils which probably belong to the Cambrian series, and are apparently newer than the Acadian or Menevian group. They consist, according to Mr. Billings, of an *Obolella, Orthisina,* and *Dictyonema,* and a trilobite of Primordial type ; and the beds holding the *Lingula* or *Obolella* are very like the Lingula shales of St. John. The series is characterized as consisting of a purple, red, and green slate, sandstones and limestones, with beds of felsite. It thus differs in character from the Acadian group, as developed at St. John, and also from the Cambrian of the Atlantic coast of Nova Scotia. It rests on the crystalline rocks of the Boisdale Hills.

Mr. Fletcher has also procured fossils from the vicinity of Miré River, where beds similar to those of St. Andrew's Channel are extensively developed, and which include an *Agnostus* and other trilobites of primordial type, but specifically distinct from those of the Acadian group ; and also a small Orthis, apparently allied to *O. Evadne,* Billings, from the Quebec group, or to *O. lenticularis,* Dalman of the British Upper Lingula flags. These fossils I regard as indicating a position probably Cambrian, but later than that of the Acadian beds of St. John. The beds containing these fossils are associated with the volcanic ash series of Southern Cape Breton. Professor Bailey informs me that in the belt of old rocks, north of the central Coal-field in New Brunswick, there are portions, apparently older than the Silurian rocks of that region, and resembling the Nova Scotia coast series, like which they are auriferous.

4. *The Ordovician* or *Siluro-Cambrian.*—In New Brunswick the band of old rocks lying on the north of the crystalline belt extending south-west from Bathurst, and composed of greenish felsites, quartzites, and slates of various kinds, is usually referred to this system. The evidence of this is, first, its appearance from under the Silurian beds in the same manner with the rocks of the Quebec group on the north ; and, secondly, the occurrence of a few characteristic fossils.* Lithologically these rocks may be regarded as corresponding somewhat closely with portions of the Quebec group, and also with the contemporaneous Skiddaw and Borrowdale series in England. According to Messrs. Bailey and Matthew, similar rocks occur also in several places in the south-west of New Brunswick,

---

* According to Prof. Bailey, *Trinucleus, Harpes, Leptaena,* and *Obolella.*

and underlie the Silurian of that region. If this view of their age is correct, then it would follow that the mixed aqueous and volcanic deposits so characteristic of the Huronian recurred in the Siluro-Cambrian, and again in the deposition of the Silurian Mascarene series.

Crossing over to Nova Scotia we have in the Cobequid Mountains a great series of slates, quartzites and volcanic rocks, evidently underlying the Silurian Wentworth series, but destitute of fossil remains. These, with their continuation in the district extending eastward from the Cobequids to the Strait of Canso and into Cape Breton, were characterized by me in 1850* as consisting of "various slates and quartzites, with syenite, greenstone, compact felspar, claystone and porphyry," and were named in "Acadian Geology" the "Cobequid group," and their age defined as intermediate between that of the lower Arisaig fossiliferous series and the Gold series (Cambrian) of the Atlantic coast. As they had afforded no fossils, and as there seemed to be a lithological and stratigraphical connection between them and the lower part of the Silurian, they were placed with that series as a downward extension, or, in part, metamorphosed members of it.

The arrangement of these rocks in the central part of the Cobequids, and also between the East River of Pictou and the east branch of the St. Mary's River, may be stated thus. There is a central mass of red intrusive syenite or syenitic granite, usually having a large predominance of red orthoclase, with a moderate quantity of hornblende and quartz. This sends veins into the overlying beds, and is itself penetrated by dykes of diabase. On this central mass rests a great thickness of felsites, porphyries, felsitic agglomerates, and diorites, evidently of volcanic origin. Upon these are gray, black, and reddish slates and quartzites, with a bed of limestone, and penetrated by metallic veins. The lower volcanic portions and the upper more strictly aqueous parts might perhaps be separated as a Lower and upper Cobequid series ; but the difference appears to depend rather on mode of deposition than on any great difference of age.

Fig. 159.—General structure of the Cobequid Range. (a) Massive Syenitic Granite. (b) Lower Cobequid Series, Felsite, Porphyry, Agglomerate, &c (c) Upper Cobequid Series ; Ferriferous Slates and Quartzite. (d) Wentworth Fossiliferous beds ; Silurian. (e) Carboniferous. (f) Triassic. (x) Veins of Syenite and Diabase.

* Journal of Geological Society, vol. vi.

Along the northern side of the Cobequids, and between Pictou and Arisaig, these beds are seen immediately to underlie the Silurian rocks, which have been disturbed with them, and are penetrated by the same igneous dykes. Dr. Honeyman appears to have observed the same relation on the Lochaber Lake and in other parts of Antigonish County. This somewhat constant association would seem so indicate that the rocks in question immediately underlie the Silurian, and are therefore themselves of Siluro-Cambrian age. On the other hand, their similarity in mineral character with the Huronian series of New Brunswick, with rocks observed in Cape Breton to rise from under Cambrian deposits, and with the Huronian rocks of Murray in Newfoundland, might induce us to assign them to an earlier date. There are, however, some differences in mineral character; as, for example, the greater prevalence of olive, black, and micaceous slates, and of highly felspathic rocks in the Cobequid series, which, while they ally this series with that of Northern New Brunswick and of the Kingston peninsula, separate it from the typical Huronian. I am therefore inclined to believe that it will ultimately be found that there are three barren series of mixed volcanic and aqueous deposits in the Acadian Provinces, separated by fossiliferous deposits, viz., (1) the Huronian, over which lie the fossiliferous Cambrian (Acadian) beds; (2) The Cobequid series, over which lie the fossiliferous Silurian; (3) The Mascarene series, belonging to the Silurian. In some districts, as in Southern New Brunswick and Cape Breton, where these series, or some of them, approach closely to each other, and are much disturbed, it may be difficult to disentangle these deposits; but I believe the distinction will be found to hold good, and will, no doubt, be facilitated by the discovery of additional fossiliferous beds.

In the meantime, I have no doubt of the identity of the greater part of the altered and volcanic beds of the hilly country extending through Pictou and Antigonish counties, and underlying the Silurian, with the Cobequid series. Further, large suites of specimens placed in my hands by Albert J. Hill, Esq., leave no room to doubt the similarity of the greater part of the rocks in the district extending from St. Peter's to Scatari in Cape Breton to the Cobequid deposits; though, as previously stated, there is reason to believe that older rocks occur both in this district and in Northern Cape Breton.

If the above views are correct, it will follow that in the Lower Silurian period, the area of Nova Scotia and New Brunswick was the theatre of extensive and long continued volcanic ejections, producing a series of rocks entirely dissimilar from those deposited at the same period in the interior continental region, though in some respects resembling those of Great Britain and those of the regions in Quebec and the United States lying east of the great Appalachian line of disturbance.

5. *The Silurian.*—In the Acadian Provinces, as in some other parts of Eastern America, the great igneous outbursts, evidenced by the masses and dykes of granite which cut the Lower Devonian rocks, make a strong line of distinction between the later and older Palæozoic. While the Carboniferous series is unaltered, except very locally, and comparatively little disturbed, and confined to the lower levels, the Silurian, and all older series, have been folded and disturbed and profoundly altered, and constitute the hilly and broken parts of the country. In the Silurian and the older periods there seems also to have been a constant mixture with the aqueous sediments in process of deposition of both acidic and basic volcanic matter, in the form of ashes and fragments, as well as probably outflows of both acidic and basic igneous rock, so that all these older formations are characterized by the presence of felsite, porphyry, petrosilicious breccia and diorite or diabase. Further, since these volcanic and tufaceous rocks, owing to their composition, are much more liable to be rendered crystalline by metamorphism than the ordinary aqueous sediments from which the bases have been leached out by water, and since they are usually not fossiliferous, the appearance is presented of crystalline non-fossiliferous rocks alternating with others holding abundant organic remains, and comparatively unaltered. The volcanic members of these series are also often very irregular in distribution, and there is little to distinguish them from each other, even when their ages may be very different. These circumstances oppose many difficulties to the classification of all the pre-Devonian rocks of Nova Scotia and New Brunswick, difficulties as yet very imperfectly overcome.

In Northern New Brunswick, at the head of the Baie de Chaleur, and more especially near Cape Bon Ami, there are Silurian limestones and shales rich in fossils, and resembling those on the Northern side of the Bay, at Gaspé and elsewhere in the Quebec region. These strata are associated with great interstratified beds of dark-coloured diabasic igneous rock. They are described in " Acadian Geology."

Messrs. Bailey and Matthew have devoted much time and labour to the rocks which crop out from under the Upper Devonian beds at Perry in Maine, and extend thence eastward into New Brunswick, where they have been named the " Mascarene series." I studied these beds in 1862, as they occur at Pigeon Hill and elsewhere near Eastport, and referred them to the Upper Silurian period,* but the tracing of their extension in New Brunswick, and the full establishment of their age, belongs to the gentlemen above named.†

These rocks are extensively developed in the south-western part of New Brunswick, and their thickness has been estimated at 2,000 feet. The following section, in ascending order, taken from the report of the Geological Survey for 1875-6, shows the general structure of the formation in Queen's County.—

Division 1. Gray clay slates, mostly of pale colour and generally somewhat calcareous. Darker gray clay slates, some of which are carbonaceous, ...about   400 feet.

Division 2. Black and dark-gray argillaceous or silicious clay slates, with very regular sedimentary bands,.............. .... about   600 "

Division 3. Dark-gray and greenish-gray earthy sandstones, the lower part compact, the upper part more slaty, greenish-gray, calcareous, or black and fissile,..... ....about   600 "

Division 4. Ash-gray and greenish-gray schistose beds, generally chloritic and calcareous, sometimes amygdaloidal and dioritic,....about   300 "

Division 5. Alternations of gray and dark-gray felsites (often porphyritic), with compact dark-gray feldspathic rock, clouded with green and purple, and with beds of dark and pale-green chloritic schist. There is a mass of felsite about 150 feet thick near the base, and a breccia conglomerate at the summit, about   800 feet or more.

These rocks, with the same general structure, are widely distributed in Southern New Brunswick, but, as might be expected, they vary in detail, more especially in the upper members. They also present a general resemblance to the belt of Silurian rocks already referred to as extending towards Bathurst, and rocks of this type are known to occur in the Silurian districts of Nova Scotia.

The fossils found in the lower members of this series near Eastport are a *Lingula* allied to *L. centrilineata* of the Lower Helderberg, and also very near to some Hamilton species, and to that found in the Lower Devonian of Gaspé, though probably different from that occurring in the Silurian of Wentworth, Pictou, and Arisaig. There are also species of *Modiomorpha*, and a species of *Loxonema*, with a small *Beyrichia* of Silurian type. Elsewhere in New Brunswick these beds have afforded species of *Strophomena*, *Orthis*, *Rhynchonella*, *Pterinea*, and corals of

---

* Paper on Precarboniferous Flora.
†Reports, Geol. Survey, 1875-6.

Silurian genera. There can thus be no doubt as to their general age, though we have not sufficient evidence to assign them to any particular horizon in the series of Silurian beds known in Nova Scotia.

The cuttings of the Intercolonial Railway have enabled Dr. Honeyman to recognize at Wentworth, on the north side of the Cobequids, the extension westward of the Silurian rocks mentioned in "Acadian Geology," and also in an earlier memoir on the metamorphic rocks of Eastern Nova Scotia,* as flanking the crystalline rocks of these hills in New Annan and Earlton. Dr. Honeyman was disposed to regard these beds at Wentworth as possibly as old as the Hudson River group ; but the fossils which I have collected in them seem to me to indicate that they are probably of the age of the Lower Arisaig series,† or about that of the Clinton of New York. They also much resemble in mineral character the Lower Arisaig beds, as well as those of similar age near Cape Gaspé, and on the Metapedia. The more characteristic fossils in my collections are :—

Graptolithus Clintonensis, Hall. Climacograpsus and Retiograpsus (?) sp. Atrypa reticularis, Dalman. Strophomena rhomboidalis, Wahl. Lingula oblonga, Hall. Orthis tenuiradiata, Hall, or allied. Orthis elegantula, Dalman, or allied. Rhynchonella neglecta, Hall, or allied. Leptocœlia intermedia, Hall, or allied. Tentaculites distans, Hall, or allied.

As usual in the shales of this series, the finer markings of the shells are not well preserved, so that it is not easy to assign them to their species. I think, however, that I cannot be wrong in referring them to the lower part of the Silurian.

At Wentworth the dark shales holding these fossils are traversed by diabase dykes, in the vicinity of which the shales have assumed a gray colour, and have been hardened so as in places to resemble felsites. It is probable that the fossiliferous beds may be unconformable to the hard slates, felsites and porphyries underlying them, but the shales have participated to some extent in the movements to which the older rocks have been exposed.

Farther eastward, at French River and Waugh's River, the representatives of the Wentworth series contain coarse limestone and hard sandstone as well as shale, but hold some of the same fossils, and at Earlton

---

* Journal Geological Society, vol. vi.

† I use the term "Lower Arisaig" in the sense attached to it in "Acadian Geology," namely, for the lower fossiliferous series of that place, in the main equivalent to the Clinton and Medina groups of New York—Llandovery of England.

loose pieces contain fossils of a somewhat higher horizon equivalent to the Upper Arisaig series.

Passing from the eastern end of the Cobequids across a bay of the Carboniferous into the Pictou area, we find well characterized Upper Silurian rocks with fossils of the Upper Arisaig (Lower Helderberg) age. These rocks have recently been somewhat carefully examined in connection with explorations of the great deposits of iron ore associated with them. It would seem that the upper half of the Upper Silurian is here quite as well developed as at Arisaig, and includes the great bed of fossiliferous hematite which is so characteristic of this region (Fig. 161). From below these beds arise thick beds of ferruginous quartzite, and of

Fig. 160.—Ideal Section, showing the general relations of the Iron Ores of the East River of Pictou. 1. Great bed of Red Hematite. 2. Vein of Specular Iron. 3. Vein of Limonite. (a) Older Slate and Quartzite series, with Felsite and Ash Rocks, etc. (b) Lower Helderberg formation and other Silurian Rocks. (c) Lower Carboniferous of the East Branch of East River.

imperfectly crystalline diorite and slaty and felsitic breccias, which would seem to be lower members of the Silurian, and which are less indurated than the rocks of similar composition previously referred to the Siluro-Cambrian and older series. These latter rocks, which also appear in the vicinity of the East River, are breccia, felsite, quartzite, slates and hydro-mica schists, which bear a close resemblance to the Cobequid series, and pass to the southward and westward of the newer rocks, no doubt forming in this region the continuation of that formation. In the central parts of the hills, at the head waters of the East River, these beds are seen, as in the Cobequids, to be invaded by great masses of an intrusive red syenite.

Eastward of the East River the continuation of the Upper Silurian rocks has been traced by Dr. Honeyman and Mr. Fletcher to Arisaig, where they were originally studied and described by the writer, and include well characterized representatives of the Clinton and Helderberg series but without the great limestones characteristic of these formations to the westward, and without distinct representatives of the Niagara group which has however been recognized in New Canaan, in Western Nova Scotia, with its characteristic fossils.*

* Canadian Record of Science, vol. ii., Acadian Geology

Mr. Matthews has recognised remains of a Placoganoid Fish akin to those of the Silurian of Britain and the United States, in rocks of this age in New Brunswick. It has been referred to the genus Pteraspis (*P. Acadica*, Matthew.)*

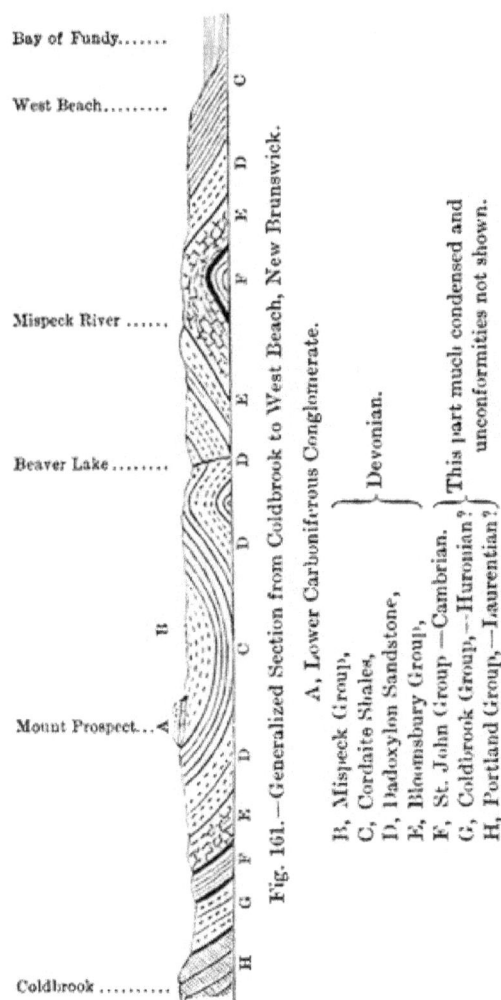

Fig. 161.—Generalized Section from Coldbrook to West Beach, New Brunswick.

A, Lower Carboniferous Conglomerate.
B, Mispeck Group,
C, Cordaite Shales,
D, Dadoxylon Sandstone,
E, Bloomsbury Group,
} Devonian.
F, St. John Group.—Cambrian.
G, Coldbrook Group,—Huronian?
H, Portland Group,—Laurentian?
} This part much condensed and unconformities not shown.

6. *The Erian or Devonian.*—The Erian System does not occupy a very wide area in the Acadian Provinces, yet, in connection with the neighboring areas in the Province of Quebec, it is of great interest, as showing perhaps more of the land life of the period, and more especially of its flora, than the Devonian of any other part of the world. In connection with this, it is to be observed that the vast development of this formation in the great Lake Erie district shows mainly its marine conditions. Yet it is satisfactory to know that Professor Hall finds erect trunks of tree-ferns and abundance of remains of fern fronds and *Psilophyton* in the Chemung sandstones of New York, and that in the marine limestones of Ohio Dr. Newberry has discovered trunks of conifers and beautifully preserved stems of tree-ferns.

* Acadian Geology and supplement, 1868 and 1878.

These rocks in the Acadian Provinces overlie the fossiliferous beds of the Lower Helderberg or Ludlow group, and underlie the Lower Carboniferous, whose peculiar flora readily distinguishes it. From these beds, thus limited, I have described or catalogued 125 species of fossil plants,[*] of which the greater part are specifically, and some generically, distinct from those of the Lower Carboniferous. A very considerable proportion of these plants have been derived from the rich plant-bearing beds near St. John, New Brunswick, so admirably explored by Messrs. Hartt and Matthew.

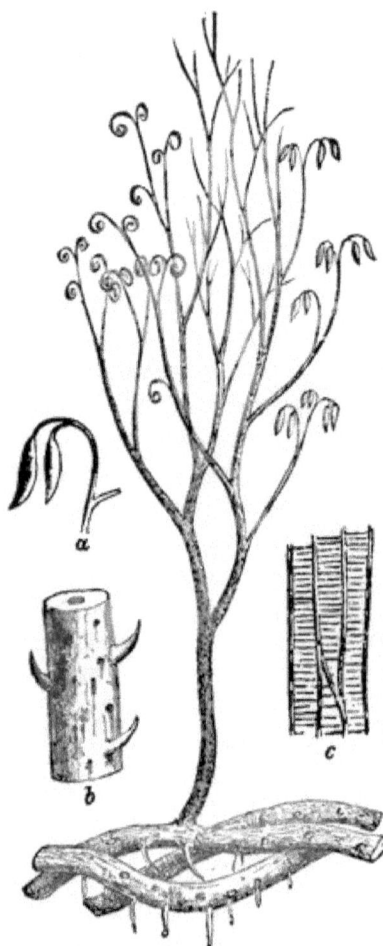

One of the most characteristic forms of the Lower Erian is the remarkable genus *Psilophyton*. The restoration given in Fig. 162 will serve to show the general character of this curious plant, which, while allied to the club-mosses in structure and habit, has remarkable peculiarities in its fructification. *Cyclopteris Jacksoni* and *C. Gaspiensis*, which should, be placed in my new genus *Archæopteris*, are characteristic of the upper members of the system. The remarkable discoveries of fossil fishes and insects in the Lower and Upper Erian of the Baie de Chaleur and St. John have been referred to in part II. The former, as described by Whiteaves, show distinctive fish faunæ of the Lower and Upper Erian, parallel with those of Europe, and corresponding to the distinct floræ of the Lower and Upper Erian.[†]

Fig. 162.—*Psilophyton princeps*, restored. *a*, Fruit, natural size ; *b*, Stem, natural size ; *c*, Scalariform tissue of the axis, highly magnified. In the restoration, one side is represented in vernation, and the other in fruit.

* Report on fossil Plants of Devonian, etc., Geol. Survey of Canada, 1871 and 1882.
† Whiteaves, Transactions Royal Society of Canada.

Fig. 163.—General Section, Minudie to Apple River. *a*, Carboniferous; *b*, Silurian. Figs. 1 to 8 indicate the Subdivisions of Logan's Section of South Joggins. In the continuation inland of the south side of the Carboniferous Synclinal are the important Coal beds of Springhill, not seen in the coast Section.

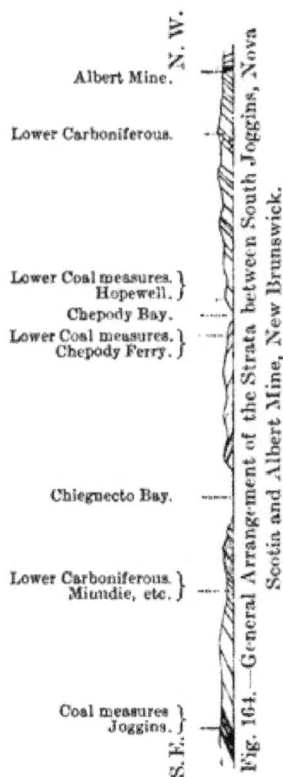

Albert Mine.

Lower Carboniferous.

Lower Coal measures. } Hopewell. }
Chepody Bay.
Lower Coal measures. } Chepody Ferry. }

Chiegnecto Bay.

Lower Carboniferous. } Minudie, etc. }

Coal measures } Joggins. }

N. W.

S. E.

Fig. 164.—General Arrangement of the Strata between South Joggins, Nova Scotia and Albert Mine, New Brunswick.

7. *The Carboniferous System.*—The total vertical thickness of the immense mass of sediment constituting the Carboniferous system in Nova Scotia may be estimated from the fact that Sir W. E. Logan has ascertained, by actual measurement at the Joggins, a thickness of 14,570 feet; and this does not include the lowest member of the series, which, if developed and exposed in that locality, would raise the aggregate to at least 16,000 feet. It is certain, however, that the thickness is very variable, and that in some districts particular members of the series are wanting, or are only slenderly developed. Still the section at the Joggins is by no means an exceptional one, since I have been obliged to assign to the Carboniferous deposits of Pictou, on the evidence of the sections exposed in that district, a thickness of about 16,000 feet; and Mr. R. Brown, of Sydney, has estimated the Coal formation of Cape Breton, exclusive of the Lower Carboniferous, at 10,000 feet in thickness.

When fully developed, the whole Carboniferous series may be arranged in four subordinate groups or formations, which are referred to in Part II, and are there taken as types of the Carboniferous system. They are (1) the *Horton Series*, or Lower Carboniferous shales and conglomerates ; (2) the *Windsor Series*, or Lower Carboniferous Limestone ; (3) the *Millstone grit* ; (4) the *Coal Formation.* Detailed features of these several members will be found in " Acadian Geology."

Fig. 165.—Sketch-map of Pictou Coal District. A, Upper Coal formation ; B, Middle Coal formation ; C, New Glasgow conglomerate ; D, Lower Carboniferous ; E, Silurian and Devonian. *Coal Areas.*—(1) General Mining Association, or Albion Mines ; (2 and 4) Acadia ; (3) Nova Scotia ; (5) Intercolonial or Bear Creek ; (6) Montreal and Pictou ; (7) German Company, and others ; (8) Sutherland's River ; (9) New Glasgow.

The area occupied by Carboniferous rocks in Nova Scotia and New Brunswick is very extensive ; and is divided by ridges of the older metamorphic rocks into portions which are considered separately in " Acadian Geology." These are—

1. The New Brunswick Carboniferous district, the largest in point of area in the Acadian provinces

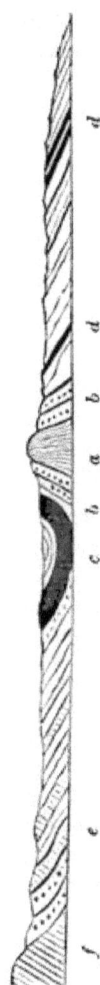

Fig. 166.—Ideal section representing the probable arrangement of the Coal formation of East River, Pictou. *a*, Devonian ridge; *b*, conglomerate; *c*, great Coal measures; *d*, minor Coal measures; *e*, Lower Carboniferous; *f*, Silurian.

2. The Cumberland Carboniferous district, bounded on the south by the Cobequid Hills, and continuous on the north-west with the great Carboniferous area of New Brunswick.

3. The Carboniferous district of Minas Basin and Cobequid Bay, and its outliers, including the long band of Carboniferous rocks extending along the south side of the Cobequids, and that reaching along the valley of the Musquodoboit River.

4. The Carboniferous district of Pictou, bounded on the south and east by metamorphic hills, and connected on the west with the Cumberland district and that last mentioned. (Fig. 165.)

5. The Carboniferous district of Antigonish county, bounded by two spurs of the metamorphic hills.

6. The narrow band of Carboniferous rocks extending from the Strait of Canseau westward through the county of Guysboro'.

7. The Carboniferous district of Richmond county and southern Inverness.

8. The Carboniferous district of Inverness and Victoria counties.

9. The Carboniferous district of Cape Breton county. (Fig. 167.)

Conditions of Deposition of the Beds.—It is evident that very various geographical conditions are implied in the deposit of this vast thickness of sediment. The Acadia of the Carboniferous period must not only have differed much from that which now is, but it must have presented very different appearances in the different portions of the Carboniferous time itself.

The conditions of deposit thus implied in the mineral character and fossils of the several formations above described, would appear to be of three leading kinds :—(1.) The deposition of coarse sediment in shallow water, with local changes leading to the alternation of clay, sand, and gravel. This predominates at the beginning of the period, recurs after the deposition of the marine limestones in the formation of the "Mill-stone-grit," and again prevails in the upper coal formation. (2.) The

growth of corals and shell-fish in deep clear water, along with the precipitation of crystalline limestone and gypsum.   These conditions
occurred during the formation of the Lower Carboniferous limestone and
its associated gypsum.   (3.) The deposition of fine sediment, and the

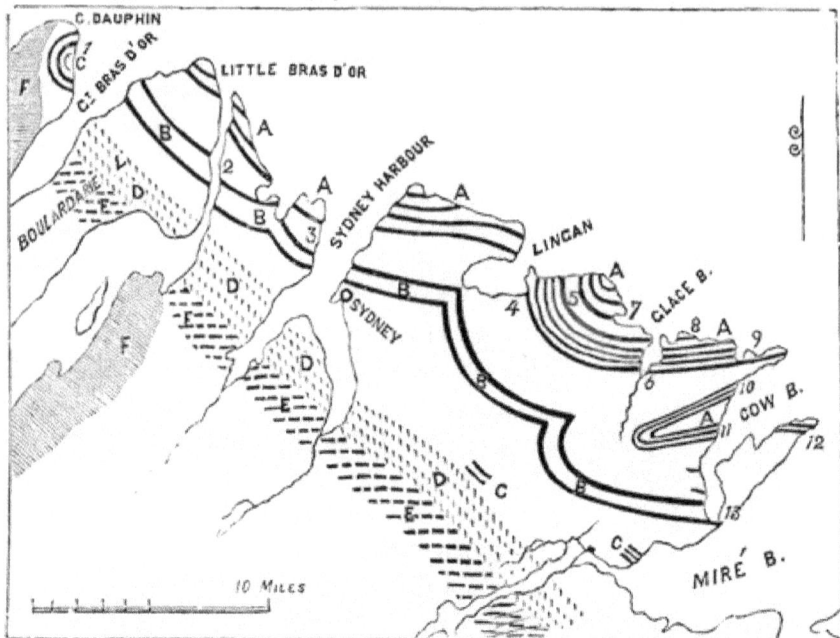

Fig. 167.—Map of Cape Breton Coal-field.   A, Upper Coal beds; B, Middle Coal
beds; C, Lowest Coal beds; D, Millstone grit; E, Lower Carboniferous;
F, Metamorphic Silurian.

1. New Campbellton Mine.	6. Caledonia Mine.	10. Block House Mine.
2. Little Bras d'Or "	7. Little Glace Bay "	11. Gowrie "
3. Sydney "	8. Clyde "	12. South Head "
4. Lingan "	9. Schooner Point "	13. Miré "
5. International "		

NOTE.—It should be observed that there are probably several beds of coal between series A and
series B, and that the lines of series B, C, D, and E are conjectural.   The town of Sydney has been placed
too far north.

accumulation of vegetable matter in beds of coal and carbonaceous and
bituminous shale, and of mixed vegetable and animal matters in the beds
of bituminous limestone and calcareo-bituminous shale.   These conditions
were those of the middle coal formation.

Within the limits of Nova Scotia and New Brunswick these conditions of deposition applied, not to a wide and uninterrupted space, but to
an area limited and traversed by bands of Silurian and Devonian rocks,

already partially metamorphosed and elevated above the sea, and along the margins of which igneous action still continued, as evidenced by the beds of trap intercalated in the Lower Carboniferous ;* while about the close of the Devonian period still more important injections and intrusions of igneous matter had occurred, as shown by the granite dykes and masses which traverse the Devonian beds, but have not penetrated the Carboniferous.† There is evidence, however, in the Carboniferous rocks of the Magdalen Islands and of Newfoundland, and in the fringes of such rocks on parts of the coast of Nova Scotia* and New England, that the area in question was only a part of a far more extensive region of Carboniferous deposition, the greater part of which is still under the waters of the Atlantic and of the Gulf of St. Lawrence.

The general phenomena of deposition above indicated, apply to all the Carboniferous areas of Nova Scotia and New Brunswick, and, so far as known, to those of the Magdalen Islands and of Newfoundland. But numerous local diversities occur, in consequence of the interference of the older elevated ridges with the regularity of deposition. In some places the entire Lower Carboniferous series seems to be represented by conglomerates and coarse sandstones. In others, the Lower Coal measures, or the marine limestones, or both, are extensively developed. These local differences are, on a small scale, of the same character with those which occur on a large scale in the northern and southern Appalachian districts and western districts of the United States, and in the different coal areas of Great Britain and Ireland, as compared with each other and with the Carboniferous districts of America. On the whole, however, it is apparent that certain grand features of similarity can be traced in the distribution of the Carboniferous rocks throughout the northern hemisphere.

It is further to be observed, that in Nova Scotia and New Brunswick, as well as in Eastern Canada, disturbances occurred at the close of the Devonian period which have caused the Carboniferous rocks to lie unconformably on those of the former ; and that in like manner the Carboniferous period was followed by similar disturbances, which have thrown the Carboniferous beds into synclinal and anticlinal bends, often very abrupt, before the deposition of the Triassic Red Sandstones. These disturbances were of a different character from the oscillations of level which occurred within the Carboniferous period. They were accom-

---

* Dawson, Quart. Journ. Geol. Soc., vol. i. 1845, p. 329.

† Dawson, Canadian Naturalist, 1860, p. 142.

panied by volcanic action, and were most intense along certain lines, and especially near the junction of the Carboniferous with the older formations.

I have noticed an apparent case of unconformability between members of the Carboniferous system near Antigonish.* In the county of Pictou, the arrangement of the beds suggests a possible unconformability of the Upper Coal formation and the Coal measures.† In New Brunswick, Prof. Bailey‡ has observed indications of local unconformability of the Coal formation with the Lower Carboniferous. But the strict conformability of all the members of the Carboniferous series in the great majority of cases, shows that these instances of unconformability are exceptional. In the section at the Joggins more especially, the whole series presents a regular dip, diminishing gradually from the margin to the middle line of the trough, where the beds become horizontal.

The most gradual and uniform oscillations of level must, however, be accompanied with irregularities of deposition and local denudation; and phenomena of this kind are abundantly manifest in the Carboniferous strata of Nova Scotia. I have described a bed in the Pictou Coal-field which seems to be an ancient shingle-beach, extending across a bay or indentation in the coast-line of the Carboniferous period. § At the Joggins, many instances occur of the sudden running out and cutting off of beds,‖ and Mr. Brown has figured a number of instances of this kind in the Coal formation of Sydney.¶ They are of such a character as to indicate the cutting action of tidal or fluviatile currents on the muddy or sandy bottom of shallow water. In some instances the layers of sand and drift-plants filling such cuts suggest the idea of tidal channels in an estuary filled with matter carried down by river-inundations. Even the beds of coal are by no means uniform when traced for considerable distances. The beds which have been mined at Pictou and the Joggins show material differences in quality and associations; and small beds may be observed to change in a remarkable manner, in their thickness and in the materials associated with them, in tracing them a few hundreds of feet from the top of the cliff to low-water mark on the beach. I have no doubt that, could we trace them over sufficiently large areas, they

* Quart. Journ. Geol. Soc., vol. i. p. 32.
† Ibid., vol. x. p. 42.
‡ "Report on Geology of Southern New Brunswick," p. 188.
§ Quart. Journ. Geol. Soc., vol. x. p. 45.
‖ Ibid., vol. x. p. 12.
¶ Ibid., vol. vi. p. 125 et seq.

would all be found to give place to sandstones, or to run out into bituminous shales and limestones, according to the undulations of the surfaces on which they were deposited, just as the peaty matter in modern swamps thins out toward banks of sand, or passes into the muck or mud of inundated flats or ponds.

8. *The Permian System.*—The Upper Coal formation was first distinguished as a separate member of the Carboniferous system in Eastern Nova Scotia by the writer, in a paper published in the first volume of the Journal of the Geological Society, in 1845—and was defined to be an upper or overlying series superimposed on the productive Coal measures, and distinguished by the absence of thick coal-seams, by the prevalence of red and gray sandstones and red shales, and by a peculiar group of vegetable fossils.

Subsequently, in my paper on the South Joggins* and in Acadian Geology, this formation was identified with the upper series of the Joggins section, Divisions 1 and 2 of Sir William Logan's sectional list, and with the Upper Barren Measures of the English Coal-fields, and the third or upper zone of Geinitz in the Coal formation of Saxony.†

Still more recently, in the "Report on the Geology of Prince Edward Island," 1871, I have referred to the upper part of the same formation, the lower series of sandstones in Prince Edward Island not previously separated from the overlying Trias.‡

In Prince Edward Island, however, where the highest beds of this series occur, they become nearly horizontal, and are overlain apparently in a conformable manner by the red sandstones of the Trias, which differ very little from them in mineral character. It thus happens that, but for the occurrence of some of the characteristic Carboniferous plants in the lower series, and of a few equally characteristic Triassic forms in the upper, it would be difficult to affirm that we have to deal with two formations so different in age.

---

* Journal Geological Society, vol. x.

† Acadian Geology, p. 149.

‡ Report on the Geological Structure of Prince Edward Island.

Fig. 168.—Bedded Igneous Rocks, at base of the Carboniferous, near McCara's Brook, Pictou County, Nova Scotia. *a,* Arisaig series, Silurian. *b,* Carboniferous Conglomerate and Sandstone. *b1,* Interstratified Trap. *b2,* Carboniferous Limestone.

In connection with this, the presumed absence of the Permian, not only here but throughout Eastern America, raises the question which I have already suggested in "Acadian Geology," whether the conditions of the Upper Coal formation may not have continued longer here than in Europe, so that rocks in the former region constituting an upward extension of the Carboniferous may synchronize with part at least of the Permian. On the one hand, there seems to be no stratigraphical break to separate these rocks from the Middle Coal formation of Nova Scotia; and their fossils are in the main identical. On the other hand, where the beds are so slightly inclined that the Trias seems conformable to the Carboniferous, no very marked break is to be expected; and some of the fossils, as the conifers of the genus *Walchia* and *Calamites gigas*, have a decided Permian tendency.

On the whole, in the Report above referred to, I declined to separate the red beds of the lower series in Prince Edward Island from the Newer Coal formation. Prof. Geinitz, however, in noticing my Report,[*] and also in a private letter, expresses the opinion that the fossils have, as an assemblage, so much of a Permian (or Dyadic) aspect that they may fairly be referred to that formation, more particularly to its lower part, the Lower-Rothliegende. This with further study of the sections on the North coast of Nova Scotia and New Brunswick induced me, in 1872, to propose the name Permo-Carboniferous for this formation.[*]

9. *The Trias.*—The great Geological Periods known as the Mesozoic and Early Tertiary are represented in the Acadian Provinces only by the Triassic system.

During all those periods in which the middle and older Tertiaries, the Cretaceous and the Oolitic systems were produced, no rocks appear to have been formed within its area, or if they were formed they have been swept away. This remark applies not only to Nova Scotia, but to an immense region extending through New Brunswick, Canada, and the Northern United States. During those long periods, these regions, thus destitute of the newer Secondary and Tertiary rocks, may have been in the interior of a great continent, or in the fathomless depths of an ocean where no sediment was being deposited; but whatever their condition, they retain no geological monuments of the lapse of time.

The distribution of the Trias indicates that, when it was deposited, the form and contour of the country already made some approach to those which it still retains. Just as the marsh mud lines the coasts of the Bay of Fundy, so do we find the Trias occupying an inner zone,

---

* Journal Geological Society, August, 1874.

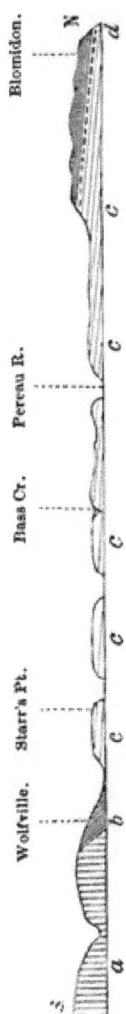

Fig. 169.—Section from Horton to Cape Blomidon, showing relation of Triassic Igneous Rocks. a, Metamorphic Slates. b, Carboniferous System. c, New Red Sandstone. d, Trap.

and appearing to have been deposited in a bay a little wider and longer than the present one. It is indeed to this bay district that, in Nova Scotia and New Brunswick, the Trias has been chiefly confined, and it may have been deposited in circumstances not very dissimilar from those of the present marshes, except that the older deposit is accompanied by evidence that active volcanoes poured out their lavas on the grandest scale in the waters and on the shores of the bay while its sandstones were being formed. While the Trias of Nova Scotia is limited to the Bay of Fundy, we have evidence in the wide extent of the same formation in Prince Edward Island, that a similar deposit was in progress in the Gulf of St. Lawrence. In the gulf, however, unlike the bay, we do not find the New Red along the coasts, but in an isolated patch separated on all sides from the continent. I may remark here, that the Trias, though patches of it are scattered over several parts of North America, is nowhere very extensive. To the southward of Nova Scotia it reappears in Connecticut, where it extends over a considerable area in the valley of the river; and in New Jersey, where another band commences that extends a great distance to the south-east, some isolated patches occurring as far south as North Carolina.

The aqueous rocks of this period in Nova Scotia and Prince Edward Island are principally coarse and soft red sandstones with a calcareous cement, which causes them to effervesce with acids, and contributes to the fertility of the soils formed from them. In the lower part of the formation there are conglomerates made up of well-worn pebbles of the harder and older rocks.

The Volcanic rocks of this period are of that character known to geologists as Doleritic Lavas and indurated tuff or volcanic ashes, and are quite analogous to the products of modern volcanoes; and, like them, consist principally of Pyroxene and Lime-felspars.

In Nova Scotia the principal areas of Triassic Red Sandstone are those around Cobequid Bay and the long, narrow valley of Cornwallis and Annapolis, celebrated for its apple orchards. The Triassic volcanic rocks form a long, narrow and elevated belt, extending westward from Cape Blomidon on the Bay of Fundy, while isolated eminences of the same rocks appear on the opposite side of the Bay.

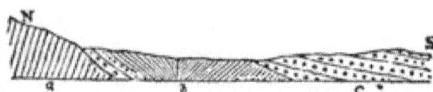

Fig. 170.—Junction of Trias and Carboniferous, Great Village River, Nova Scotia. a, Siluro-Cambrian. b, Carboniferous. c, Trias.

The beds of the Triassic series, as seen in Prince Edward Island, and more especially in its northern part, consist chiefly of soft red sandstone, with some buff coloured beds and red and mottled clays. Associated with them are conglomerates and hard calcareous and concretionary sand-stones, passing into bands of arenaceous limestone, which is in some places a dolomite. They present a division into an upper and lower group which may be held to represent respectively the Bunter and Keuper Sandstones, the lower member containing a greater proportion of gray, purplish and pebbly beds.

The dips are so low, and the beds so much affected by oblique stratifica-tion, that those of the Trias cannot be said to be unconformable to the underlying Carboniferous rocks ; and for this reason, as well as on account of the similarity in mineral character between the two groups, some uncertainty may rest on the position of the line of separation. That above stated depends on fossils, or a somewhat abrupt change of mineral character, and on a slight change in the direction of the dip. These beds spread over much of the island, though with no great thickness.

Fossils are rare in the Triassic beds. Of plants, one of the most interesting is a species of coniferous tree distinct from that occurring in the Carboniferous beds beneath, and allied to *Dadoxylon Keuperianum* of the European Trias. I have described it under the name *D. Edvard-ianum.* Another is apparently a small cycadean stem, which I have de-scribed as *Cycadeoidea (Mantellia) Abequidensis,* from the old Micmac name of the Island.* Besides these there are Knorria-like stems, a coarsely marked *Sternbergia,* and impressions resembling fucoids. The only animal fossil yet known is *Bathygnathus borealis,* Leidy, a member of the group of carnivorous dinosaurs, the highest known reptiles, and an order very characteristic of the Mesozoic.

* Report p. 45.

10. *The Pleistocene.*—After a great gap in the Geological succession, the Maritime Provinces present a moderate development of the clays, sands and gravels of the Pleistocene or Glacial age, resting on striated rock surfaces. These I have all along regarded as evidence of arctic and sub-arctic conditions, but not of land glaciation within the area of the Maritime Provinces, except on the flanks of the higher mountain ridges.

I fail to find, either in the Acadian Provinces or in Canada proper, any indication of a great continental glacier. There is evidence of great depression of the land, accompanied with a reduction of the mean temperature to such an extent that the hills remaining above water were occupied with local glaciers, and formed areas of denudation, while the lower lands, traversed by northern currents of ice-cold water, bore floating ice throughout the year, and this was steadily pushed by the lower currents from the north-east; while in periods of extreme submergence there was a drift, perhaps caused by prevailing winds, from the north-west. In these circumstances the boulder clay and the lower part of the Leda clay were formed, and are consequently non-fossiliferous, or hold only a few Arctic shells. Re-elevation brought shallowness, and consequently warmer water, and eventually land surfaces, and introduced the modern climate.

Whatever the cause of this submergence, the fact of its occurrence is proved by the marine clays and the high-level sea beaches. Mr. Richardson of the Geological Survey has found these terraces 1225 feet above the sea on the coast of Newfoundland, and the evidence of travelled stones would take the sea to the tops of the highest hills in Eastern America, 6,000 feet above the sea. The drift phenomena of the western plains and the Rocky Mountains imply subsidence there to the extent of, at least, 4,400 ft.* Now, the existing climates of the North Atlantic, as compared with those of the Post-pliocene, point precisely to the natural effects of such a submergence, while the action of local glaciers, of pack and pan ice, and of drifting bergs, as now actually observed, would, if intensified, as they must have been by the causes supposed, give all the observed effects of glaciation. There is therefore no geological necessity to appeal to the varying eccentricity of the earth's orbit and the precession of the equinoxes, or to an imagined change of the earth's axis of rotation, or of the obliquity of the ecliptic, or of the energy of the sun's radiation. If these, or any of them, can be proved on other grounds, geologists may fairly be called on to allow for their

* G. M. Dawson, Report on 49th parallel.

influence ; but there is no geological necessity for them, other than the exigencies of an imaginary period or succession of periods of continental glaciation, of which unquestionably there is no geological evidence in Eastern America. For facts in support of this view, I may refer not only to the chapter on Acadian Geology on this subject, but to my subsequently published Notes on the Post-pliocene of Canada.*

I may, however, quote the explanation which I gave in the first edition of Acadian Geology, in 1885, and to which I still adhere :—

"In reasoning on this subject as regards Nova Scotia, I have the advantage of appealing to causes now in operation within the country, and which are at present admitted by the greater number of modern geological authorities to afford the best explanation of the phenomena. In the first place, it may at once be admitted that no such operations as these which formed the drift are now in progress on the surface of the land, so that the drift is a relic of a past state of things, in so far at least as regards the localities in which it now rests. In the next place, we find, on examining the drift, that it strongly resembles, though on a greater scale, the effects now produced by frost and floating ice. Frost breaks up the surface of the most solid rocks, and throws down cliffs and precipices. Floating ice annually takes up and removes immense quantities of loose stones from the shores, and deposits them in the bottom of the sea, or on distant parts of the coasts. Very heavy masses are removed in this way. I have seen in the Strait of Canseau large stones, ten feet in diameter, that had been taken from below low-water mark and

Fig. 171.—Modern Travelled Stone, Petitcodiac River.

pushed up upon the beach. Stones so large that they had to be removed by blasting, have been taken from the base of the cliffs at the Joggins and deposited off the coal-loading pier, and I have seen resting on the mud-flats at the mouth of the Petitcodiac River a boulder at least eight feet in length, that had been floated by the ice down the river (Fig. 171.) Another testimony to the same fact is furnished by the rapidity with

* Mr. Chalmers has more recently investigated the facts as to local drift in the Baie des Chaleurs and the mountains south of the Lower St. Lawrence.—Reports, Geological Survey of Canada, and Can. Record of Science, 1889.

which huge piles of fallen rock are removed by the floating ice from the base of the trap cliffs of the Bay of Fundy. Let us suppose, then, the surface of the land, while its projecting rocks were still uncovered by surface deposits, exposed for many successive centuries to the action of alternate frosts and thaws, the whole of the untravelled drift might have been accumulated on its surface. Let it then be submerged until its hill-tops should become islands or reefs of rocks in a sea loaded in winter and spring with drift-ice, floated along by currents, which, like the present Arctic current, would set from N.E. to S.W. with various modifications produced by local causes. We have in these causes (along with the action of local glaciers) ample means for accounting for the whole of the appearances, including the travelled blocks and the scratched and polished rock-surfaces. This, however, is only a general explanation. Had we time to follow it into details, many most interesting and complicated facts and processes would be discovered." *

Glacial striation is very frequent wherever fresh surfaces of rock are exposed. The following are instances of its direction :—

Point Pleasant, and other places near Halifax, exposure south, very distinct and beautiful striation.......	S. 20° E. to S. 30° E.
Head of the Basin, exposure south, but in a valley,......	E. & W. nearly.
La Havre River, exposure S.E.......	S. 20° W.
Petite River, exposure S.E.............................	S. 20° E.
Bear River, exposure N	S. 30° E.
Rawdon, exposure N......	S. 25° E.
The Gore Mountain, exposure N., two sets of striæ, respectively	S. 65° E. & S. 20° E.
Windsor Road, exposure not noted	S.S.E.
Gay's River, exposure N	Nearly S. & N.
Musquodoboit Harbour, exposure S.	Nearly S. & N.
Near Pictou, exposure E., in a valley.........................	Nearly E. & W.
Polson's Lake, summit of a ridge............................	Nearly N. & S.
Near Guysboro', exposure not noted...	Nearly S. & N.
Sydney Mines, Cape Breton, exposure S.......................	S. 30° W.†

The above instances show a tendency to a southerly and south-easterly direction, which accords with the prevailing course in most parts of North-eastern America. Local circumstances have, however, modified this prevailing direction; and it is interesting to observe that while S.E. is the prevailing direction in Acadia and New England, it is exceptional in the St. Lawrence valley, where the prevailing direction is S.W.‡

---

* See " Notes on Canadian Pleistocene," Canadian Naturalist.

† The above and other courses in this volume are *magnetic*, the average variation being about 18° W.

‡ Logan, " Report on Geology of Canada."

Professor Hind has given a table of similar striation in New Brunswick, showing that the direction ranges from N. 10° W. to N. 30° E., in all except a very few cases. On Blue Mountains, 1650 feet above the sea, it is stated to be N. and S. As in Nova Scotia, N.W. and S.E. seems to be the prevailing course.

The occurrence of Laurentian boulders from Labrador in Nova Scotia, shows the great distance of transport in some cases, and its direction to the south-west, while there are abundant examples of local driftage, more especially from the higher lands and along the lines of depression of the surface, and some of this is no doubt the work of local glaciers in intervals of land elevation.*

The Pleistocene of Prince Edward Island may be taken as a good illustration of the deposits of this age.

The Triassic and Upper Carboniferous rocks of this island consist almost entirely of red sandstones, and the country is low and undulating, its highest eminences not exceeding 400 feet. The prevalent Post-pliocene deposit is a boulder clay, or in some places boulder loam, composed of red sand and clay derived from the waste of the red sandstones. This is filled with boulders of red sandstone derived from the harder beds. They are more or less rounded, often glaciated, with striæ in the direction of their longer axis, and sometimes polished in a remarkable manner, when the softness and coarse character of the rock are considered. This polishing must have been effected by rubbing with the sand and loam in which they are imbedded. These boulders are not usually large, though some were seen as much as five feet in length. The boulders in this deposit are almost universally of the native rock, and must have been produced by the grinding of ice on the outcrops of the harder beds. In the eastern and middle portion of the island, only these native rocks were seen in the clay, with the exception of pebbles of quartzite, which may have been derived from the Triassic conglomerates. At Campbellton, in the western part of the island, I observed a bed of boulder clay filled with boulders of metamorphic rocks similar to those of the mainland of New Brunswick. Striæ were seen only in one place on the north-eastern coast and at another on the south-western. In the former case their direction was nearly S.W. and N.E. In the latter it was S.70° E.

No marine remains were observed in the boulder clay; but at Campbellton, above the boulder clay already mentioned, there is a limited area occupied with the beds of stratified sand and gravel, at an elevation

* See Chalmers, l.c., and Geological Magazine, 1889.

of about 50 feet above the sea, and in one of the beds there are shells of *Tellina Grœnlandica*.

On the surface of the country, more especially in the western part of the island, there are numerous travelled boulders, sometimes of considerable size. As these do not appear in situ in the boulder clay, they may be supposed to belong to a second or newer boulder drift, similar to that which we find to be connected with the Saxicava sand in Canada. These boulders being of rocks foreign to Prince Edward Island, the question of their source becomes an interesting one. With reference to this, it may be stated in general terms that the majority are granite, syenite, diorite, felsite, porphyry, quartzite, and coarse slates, all identical in mineral character with those which occur in the metamorphic districts of Nova Scotia and New Brunswick, at distances of from 50 to 200 miles to the south and south-west, though some of them may have been derived from Cape Breton on the east. It is further to be observed, that these boulders are most abundant and the evidences of denudation of the Trias greatest in that part of the island which is opposite the deep break between the hills of Nova Scotia and New Brunswick, occupied by the Bay of Fundy, Chignecto Bay, and the low country extending thence to Northumberland Strait, an evidence that the boulder drift was connected with currents of water passing up this depression from the south or south-west. Similar local drift occurs in Nova Scotia, though there the predominant direction is from the northward.

Besides these boulders, however, there are others of a different character; such as gneiss, hornblende-schist, anorthosite and Labradorite rock, which must have been derived from the Laurentian rocks of Labrador and Canada, distant 250 miles or more to the northward. These Laurentian rocks are chiefly found on the north side of the island, as if at the time of their arrival the island formed a shoal, at the north side of which the ice carrying the boulders grounded and melted away. With reference to these boulders, it is to be observed that a depression of four or five hundred feet would open a clear passage for the Arctic current entering the Straits of Belle Isle to the Bay of Fundy; and that heavy ice carried by this current would then ground on Prince Edward Island, or be carried across it to the southward. If the Laurentian boulders came in this way, their source is probably 400 miles distant in the Strait of Belle Isle. On the north shore of Prince Edward Island, except where occupied by sand dunes, the beach shows great numbers of pebbles and small boulders of Laurentian rocks. These are said by the inhabitants to be cast up by the sea or pushed up by the ice in spring.

Whether they are now being drifted by ice direct from the Labrador coast, or are old drift being washed up from the bottom of the gulf, which north of the island is very shallow, does not appear. They are all much rounded by the waves, differing in this respect from the majority of the boulders found inland. I may add here that Laurentian boulders have been observed on the north shore of Nova Scotia.* Dr. Honeyman records their appearance even on the Atlantic coast.

The older boulder clay of Prince Edward Island, with native boulders, must have been produced under circumstances of powerful ice action, in which comparatively little transport of material from a distance occurred. If we attribute this to a glacier, then as Prince Edward Island is merely a slightly raised portion of the bottom of the Gulf of St. Lawrence, this can have been no other than a gigantic mass of ice filling the whole basin of the gulf, and without any slope to give it movement except towards the centre of this great though shallow depression. On the other hand, if we attribute the boulder clay to floating ice, it must have been produced at a time when numerous heavy bergs were disengaged from what of Labrador was above water, and when this was too thorough-ly enveloped in snow and ice to afford many travelled stones. Further, that this boulder-clay is a sub-marine and not a subaerial deposit, seems to be rendered probable by the circumstance, that many of the boulders of the native sandstone are so soft that they crumble immediately when exposed to the weather and frost.

The travelled boulders lying on the surface of the boulder clay evidently belong to a later period, when the hills of Labrador and Nova Scotia were above water, though lower than at present, and were sufficiently bare to furnish large supplies of stones to coast ice carried by the tidal currents sweeping up the coast, or by the Arctic current from the north, and deposited on the surface of Prince Edward Island, then a shallow sand-bank. The sands with sea-shells probably belonged to this period, or perhaps to the later part of it, when the land was gradual-ly rising. Prince Edward Island thus appears to have received boulders from both sides of the Gulf of St. Lawrence during the later Post-pliocene period; but the greater number from the south side, perhaps because nearer to it. It thus furnishes a remarkable illustration of the transport of travelled stones at this period in different directions; and in the comparative absence of travelled stones in the lower boulder clay; it furnishes a similar illustration of the homogeneous and untravelled char-

---

* Notes on Post-pliocene, 1872. p. 112.

acter of that deposit, in circumstances where the theory of floating ice serves to account for it at least as well as that of land ice, and, in my judgment, greatly better.

Subdivisions of the Pleistocene Deposits.—In Acadian Geology, and in my Memoirs on the Pleistocene of the St. Lawrence Valley, I have proposed a three-fold division of these beds into *Boulder clay, Leda clay,* and *Saxicava sand* and gravel, to which may be added the old peaty deposit observed under the boulder clay in Cape Breton. Mr. Matthew has since recognized in New Brunswick certain beds only locally developed in the St. Lawrence Valley, and which I have been hitherto disposed to regard as depending on the action of streams from the land or littoral agencies, but which he regards as marine deposits. They are gravels and sands underlying the boulder clay, and as yet destitute of fossils.

The complete series of Pleistocene beds in Acadia and Canada would thus stand as follows, in ascending order, though it is to be observed that the whole series is not to be found developed at any one place :—

(a.) Peaty terrestrial surface anterior to boulder clay.

(b.) Lower stratified gravels—(Syrtensian deposits of Matthew).

(c.) Boulder clay and unstratified sands with boulders. Fauna, when present, extremely Arctic.

(d.) Lower Leda clay, with a limited number of highly Arctic shells, such as are now found only in permanently ice-laden seas.

(e.) Upper Leda clay and sand, or Uddevalla beds, holding many sub-Arctic or boreal shells similar to those of the Labrador coast at present.

(f.) Saxicava sand and gravel, either non-fossiliferous, or with a few littoral shells of boreal or Acadian types.

This table may be regarded as giving a complete statement of the series of deposits in the Post-pliocene, not only in the Acadian Provinces, but throughout North-eastern America.

Fossils of the Pleistocene.—Near the city of St. John, are gray and reddish clays, holding fossils which indicate moderately deep water, and are, as to species, identical with those occurring in similar deposits in Canada and in Maine. They would indicate a somewhat lower temperature than that of the waters of the Bay of Fundy at present, or about that of the northern part of the Gulf of St. Lawrence. They correspond to the Leda clay of Canada and Maine.

Mr. C. F. Hartt has given, in Prof. Bailey's Report on New Brunswick, the following list of fossils from these beds. I have affixed an asterisk to the species found also in the Leda clay and Saxicava sand of the Province of Quebec.—

Articulata.

  Balanus Hameri,* *Asc.*, Lawlor's Lake.
  B. crenatus,*   "   "

Mollusca.

  Pecten islandicus,* *Linn.*, Lawlor's Lake, R.R. Depot, Saint John.
  P. tenuistriatus, Migh.,   "   "   "   "
  Mytilus edulis, *Linn.*   "   "   "   "
  Cardium pinnulatum, *Con.*   "   "   "   "
  Tellina Grœnlandica* (=T. Balthica *Linn.*), Lawlor's Lake, etc.
  T. calcarea* (= Macoma sabulosa, *Stp.*), Duck Cove, etc.
  Leda Jacksoni (= L. pernula*), Lawlor's Lake.
  L. truncata,* Duck Cove ; Lawlor's Lake ; R.R. Depot, Saint John.
  Nucula antiqua (var. of N. tenuis)*   "   "
  Mya arenaria.*   "   "
  M. truncata,*   "   "
  Aphrodite (Serripes) Grœnlandica, *Beck*, Duck Cove, etc.
  Cardium islandicum,* *Linn.*
  Mesodesma, R.R. Depot.
  Saxicava distorta, *Say* (= S. rugosa, *Linn.*)*
  Lyonsia arenosa, Duck Cove.*
  Lacuna neritoidea,* *Gould*, Duck Cove.
  Pandora trilineata,   "
  Natica clausa, *Sow,*   "
  Buccinum undatum,* *Linn,*   "

Bryozoa.—several species undetermined, Taylor's Island, Lawlor's Lake, etc.

Radiata.

  Ophioglypha Sarsii, *Lutk.*, Saint John, Duck Cove.*
  Toxopneustes drobachiensis, (Echinus granulatus, *Say.*),* Red Head, Lawlor's
      Lake.

Plants.—Algæ, three species, undetermined.—Manawagonis.

In some specimens sent to me by Mr. Matthew, I find in addition to the forms above enumerated, some microscopic organisms, more especially *Polystomella striato-punctata* (umbilicata of Walker), and several species of *Cythere* ; and among the Bryozoa I recognize *Pustulipora*, *Tubulipora serpens* and *Crisia eburnea*, all in small fragments.

The Rev. Mr. Paisley has published in the "Canadian Naturalist" (1872) a list of shells obtained from a railway cutting on the Tattagouche River, near Bathurst, in New Brunswick. They were found in beds of Leda clay passing upwards into sand and gravel. At the Jacquet River in the same district, the bones of a small cetacean have been found, and have been described by Dr. Gilpin and Dr. Honeyman.* They are

* Trans. N.S. Institute.

referred by Dr. Gilpin to *Beluga Vermontana* of Thompson from the Pleistocene of Vermont. Similar bones have been found in the Leda clay of the St. Lawrence Valley, and have been compared by the late Mr. Billings with the skeleton of the recent *B. catodon*, L., of the St. Lawrence, with which the so-called *B. Vermontana* is probably identical, as the specimens above referred to, and examined by Billings, certainly were.

Mr. Matthew has noticed the occurrence of *Tellina Grœnlandica* at Horton Bluff, and he makes the important observation that the shells found on the coast of the Baie de Chaleur are of more modern type than those in the Bay of Fundy, which conform more nearly to the assemblage found in these deposits on the New England coasts, so that the existing geographical regions were already to some extent established on the coast of North America in the period of the Upper Leda clay.

Just as we attribute the formation of the older or boulder drift to the action of water and ice, while the land was subsiding beneath a frozen sea, so we may assign as the cause of the superficial gravels the action of these same waters while the country was being elevated above their level. Many of the mounds of gravel have evidently been formed by currents of water rushing through and scooping out the present valleys. Some of the more regular ridges are apparently of the nature of the gravel beaches which are thrown by the sea across the mouths of bays and coves, and may mark the continuance of the sea-level unchanged for some time in the process of elevation. Others may have been pressed up by the edges of sheets of ice, in the manner of the ridges along the borders of our present lakes. That the action of ice in some form had not ceased, we have evidence in the large boulders sometimes found on the summits of the gravel ridges.

In the island of Cape Breton the bones of a large elephantine quadruped, *Mastodon Americanus*, have been found in connection with the superficial gravel, which may be regarded as of Post-glacial or late Glacial age. This gigantic creature probably inhabited our country at the close of the Glacial period, and may have been contemporary with some of the present animals, but possibly extinct before the introduction of man, though the Micmac Indians seem to have had traditions of its existence. In Cape Breton the animal must have attained to its usual great dimensions, for a thigh-bone, now in the Provincial Museum in Halifax, though apparently somewhat worn, measures three feet eleven inches in length.

11. *The Modern Period.*—Changes of Level.—In the surveys of the Baie Verte Canal, made by Mr. Page under authority of the Dominion Government, I find it stated in the report of Mr. Baillargé, that between the Missaquash River and Cumberland Creek to the north of the point where I have discovered the remarkable submarine forest of Fort Lawrence,[†] stumps of trees were seen rooted in earth for more than half a mile along the shore, and extending from low water mark to the bank. They are stated to be from 32.8 feet to 22.3 feet below the level of the highest tides. The surveyors recognized spruce, beech, pine and tamarac, all in a fair state of preservation, and rooted in a vegetable mould underlaid by a sandy subsoil. The bed of stumps *in situ* at Fort Lawrence, noticed in part II., page 122, and fully described in my " Acadian geology," is a still more interesting example. In my Report on Prince Edward Island I have noticed evidence of similar modern subsidence, though to a less amount. These facts place themselves in connection with the probability that in America, as in Europe, a period of continental elevation succeeded the great Pleistocene subsidence, and has been followed by a depression in more modern times. This consideration seems to account for some otherwise anomalous facts in connection with the distribution of modern marine animals.

Note on the Correlation of the Geology of the Maritime Provinces with that of Europe.[*]

As early as 1855, in the first edition of Acadian Geology, the author had indicated the close resemblance in structure and mineral productions of Nova Scotia and New Brunswick with the British Islands, and in subsequent editions of the same work, further illustrations were given of this fact. Recent researches by Bailey, Matthew, Fletcher, Ells and others, still more distinctively indicate this resemblance, as well as the distinctness of the Maritime Geology from that of the great interior plateau of Canada and the United States. In short, the Geology of the Atlantic margins of America and Europe is substantially the same, and distinct from that found west of the Apalachians in America and in Central and Eastern Europe. In this fact has originated much of the difficulty experienced in correlating the geological formations of Eastern Canada with those of Ontario, of New York and Ohio, as well as similar difficulties in Europe, which have led to much controversy and difference of

---

* Abstract of a paper in the Journal of the Geological Society of London, 1888.

classification and nomenclature. The system of Palæozoic sediments, employed for the interior plateau of the American continent, thus requires very important modifications when applied to the Atlantic coast, and neglect of this has led to serious misconceptions.

The rugged islands of Laurentian and Huronian rocks correspond on both sides of the Atlantic, and show an identity of succession in deposits as well as a synchronism of the great folds and lateral pressures which have disturbed these old formations. The Cambrian sediments and fossils as originally described by Hartt, and more recently and in so great detail by Matthew, are in close correspondence with those of Wales and not identical with those of internal America. The recent paper of Lapworth on the Graptolites affords evidence of the same kind, and shows that these were Atlantic animals in their time. It also throws much additional light on the Quebec group of Logan considered as an Atlantic marginal formation, representing a great lapse of time in the Cambrian and Ordovician periods. The author has long ago shown that the Siluro-Cambrian or Ordovician of Nova Scotia conformed more nearly to that of Cumberland and Wales than to the great limestone formations of Quebec, Ontario and New York. The Upper Silurian also is of the type of that of England and Wales, a fact very marked in its fossil remains as well as in its sediments.

The parallelism in the Erian or Devonian in both countries is most marked, both in rocks and fossils, and while this is apparent in the fishes, as worked up by Mr. Whiteaves, it is no less manifest in the fossil plants as described by the author.

The Carboniferous, in its limited troughs, the character of its beds and its fossil animals and plants, also points to a closer relationship in that period between the two shores of the Atlantic than between the Atlantic coast and the inland area.

The Trias of Nova Scotia and of Prince Edward Island, as the author showed in 1868,* resembles that of England very closely, in its aqueous deposits and in its associated trappean rocks.

Beyond this the Geology of the Maritime Provinces presents no materials for comparison till we arrive at the boulder drift and other pleistocene deposits. In regard to these, without entering into disputed questions any farther than to say that the observations of the author, as well as those more recently made by Mr. Chalmers, conclusively prove that submergence and local ice-drift were dominant as causes of distribution of boulders and other material, there was evidence of great similarity.

* Journal of Geol. Soc. of London, vol. VI.

The marine beds described by Mr. Matthew at St. John are precise equivalents of the Clyde beds of Scotland, as are the upper shell-bearing beds of Prince Edward Island and Baie des Chaleurs of those in Aberdeenshire and other parts of Scotland, and the Uddevalla beds of Sweden. The boulders drifted from Labrador to Nova Scotia were the representatives of those in Europe scattered southward from Scandinavia, and the local drift in various directions from the hills was the counterpart of that observed in Great Britain. The survival of *Mastodon Americanus* in Cape Breton, to the close of the Pleistocene, is a decided American feature, and so is the absence of any evidence of Pleistocene man.

In so far, therefore, as palæontology and the subdivisions of systems of formations is concerned, the geology of the Maritime Provinces is European, or perhaps more properly Atlantic, rather than American, and is to be correlated rather with the British Islands and Scandinavia than with interior Canada and the United States. The latter country, even on its eastern coast, possesses a much less perfect representation of these Atlantic deposits than that in the Maritime Provinces and Newfoundland, though the recent studies of Crosby, Dale and others, are developing new points of this kind in the geology of New England, and Hitchcock and others have shown that the New Brunswick Geology extends in Maine.

Conspectus of Geological Formations in the Acadian Region, with some typical localities, described in "Acadian Geology" or in Reports of Geological Survey of Canada.—

Pleistocene.	Saxicava Sand and Gravels.	Generally distributed.
	Leda Clay.	Near St. John, N. B.
	Boulder Clay or Till.	Generally distributed.
Triassic.	Upper Red Sandstone and Traps of Bay of Fundy. Upper Red Sandstones of P. E. I.	Coast of Basin of Minas and N. coast of Prince Edward Island.
Carboniferous.	Upper Carb. and Permo-Carb.	N. Pictou and coast of P. E. I.
	Middle Carboniferous.	Joggins, Pictou, Cape Breton, etc.
	Millstone Grit.	Pictou, Colchester, etc.
	Windsor Group. (Limestone, Gypsum, etc.)	Windsor, Shubenacadie R., etc.
	Horton Group. (Lower Coal Measures.)	Horton Bluff, Hillsborough, etc.
Devonian.	Catskill. { Scaumenac Beds (Baie des Chaleurs).	Scaumenac Bay, Gaspé Bay.
	Chemung and Portage.	Mispec, S. New Brunswick.
	Hamilton. { St. John Series. (Cordaite Shale. Dadoxylon Sandstones.)	Courtney Bay, near St. John, N.B.
	Oriskany. Nictau Series.	Nictau, Nova Scotia.
Silurian.	Lower Helderberg Upper Arisaig Series.	Arisaig and East River, Pictou.
	Niagara. New Canaan Series.	New Canaan, S.W. New Brunswick.
	Clinton. Lower Arisaig Series.	Arisaig.
[Si-Cambr'n.]	Cobequid Series?	Cobequid Mts., East R., Pictou.
	Graptolitic Shales of New Brunswick.	N. New Brunswick.
Cambrian.	Upper Cambrian. { Miré and St Andrew Series, Cape Breton.	Southern Cape Breton.
	Middle Cambrian. { Acadian Series	St. John, New Brunswick. Atlantic coast, Canseau to Yarmouth.
	Lower Cambrian. { Atlantic Coast Series, Nova Scotia.	Basal series, near St. John, N.B.
Huronian.	Felsitic, Chloritic, and Epidotic Rocks of St. John, Yarmouth, and Cape Breton, in part	St. John, N.B. and eastward, Boisdale Hills, C.B.
Laurentian.	Gneiss, Quartzite and Limestone of St. John and St. Anne's Mountain, Cape Breton.	Southern New Brunswick and Northern Cape Breton.

(In Devonian section, spanning note: Gaspé Sandstones and equivalents on Baie des Chaleurs.)

(In Carboniferous section, spanning note: Bonaventure formation (in N. B. New Brunswick & E. Quebec.))

II. THE CANADIAN REGION PROPER.

(Quebec and Ontario.)

This region in its eastern part consists of the Siluro-Cambrian and Silurian valley of the Lower St. Lawrence, extending from Anticosti and Gaspé to the Thousand Islands, with portions of the folded rocks of the Apalachians in the Eastern Townships of Quebec in the south, and of the great Laurentian nucleus of the North American continent in the north. These features belong mainly to the Province of Quebec. In the western or Ontario section it consists of the northern margin of the great palæozoic plateau of the interior of North America along with a wide area of Huronian and Laurentian country to the northward and westward.

We thus have in the Provinces of Quebec and Ontario the following geological districts :—

1. The southern portion of the great Eozoic (Laurentian and Huronian) nucleus of Northern Canada, separating the valley of the St. Lawrence and the Great Lakes from the Arctic Basin.

2. The hilly and broken region of Siluro-Cambrian and Cambrian and Pre-Cambrian rocks extending through the south-eastern part of Quebec from Gaspé to the United States boundary, and constituting a north-eastern extension of the Apalachian and Green mountain ranges.

3. The Siluro-Cambrian plain of the Lower St. Lawrence, east of the Thousand Islands, and occupied principally by slightly inclined formations ranging from the Potsdam sandstone to the Hudson River series, inclusive, with some limited areas of Silurian beds and occasional masses of igneous rock of Silurian date.

4. The Palæozoic plain of Ontario, consisting of nearly horizontal beds, of ages ranging from the Potsdam to the Upper Erian inclusive.

As these districts have in Part II. been taken as types of the Eozoic and older Palæozoic Periods, and as they have been described with much local detail in the "Geology of Canada" by Sir W. E. Logan, and in Reports of the Geological Survey, it will not be necessary to treat of them very fully in this place, but their great leading features will be given, with reference to the above mentioned works for details.

1. The Archean or Eozoic district occupies the north shore of the Gulf and River St. Lawrence, with a few marginal patches of Siluro-Cambrian, from the Straits of Belleisle to Quebec. Below Quebec the margin of the Laurentian area begins to recede from the St. Lawrence and to leave a gradually widening band of flat Siluro-Cambrian rocks between the river and the Laurentian hills as far as the confluence of the Ottawa

river. It then follows the valley of the Ottawa to the westward. About 100 miles west of the confluence of the Ottawa with the St. Lawrence it is suddenly deflected to the south-east, and crossing the St. Lawrence in a low and narrow band at the Thousand Islands, connects as by an isthmus the main Laurentian district with the great peninsular mass of the Adirondack mountains in New York, in which the Laurentian rocks attain their greatest altitude, Mt. Marcy being 5,400 feet in height. West of the Thousand Islands the southern boundary of the Laurentian and Huronian strikes westwardly across Ontario, till it reaches the Georgian Bay of Lake Huron, whence it sweeps in a bold curve, bordered in part by Cambrian and Huronian rocks, around the north shore of Lake Superior. Beyond the western end of that great lake it turns to the north-west and skirting the west side of Lake Winnipeg runs into the Arctic region. In this district are the most important deposits of Magnetite, Hematite, Apatite and Graphite, mined in Quebec and Ontario, and also valuable quarries of granite, marble and mica, and deposits of copper and silver. The Laurentian area is on Georgian Bay and the North shore of Lake Superior, fringed and partly covered with Huronian and Kewenian rocks. The typical Huronian of Logan is that on the north of Georgian Bay. (See Part II.)

2. The second district, that of the Eastern Townships of Quebec and thence to Gaspé, is principally characterized by the prevalence of the Quebec group of Sir William Logan, consisting of black, gray and red shales, sandstones, and coarse limestone conglomerates, with some bands of limestone and dolomite. Originally a peculiar sub-marginal formation belonging to the Atlantic basin, these beds have been greatly folded and disturbed and broken off by faults from the undisturbed plateau formations to the westward. Their age, as indicated by fossils, ranges from that of the Potsdam to the Chazy, though the greater part would seem to belong to the lower part of the Siluro-Cambrian and to be equivalent to the Upper Calciferous and Chazy of the inland plateau.

As the question of the special character of the Quebec group is of importance to the understanding of the geology of the whole region, it may be useful to give from a note contributed by the author to Harrington's Biography of Sir William Logan (1883) the following explanation, with ideal sections intended to illustrate it.—

There seems to have occurred, before the Siluro-Cambrian age, an elevation of the Laurentian nucleus of the American Continent which caused either an absence of deposit or very shallow water deposits over the greater part of the area of North America, while thick deposits were laid down on the Atlantic border. This difference and its causes are

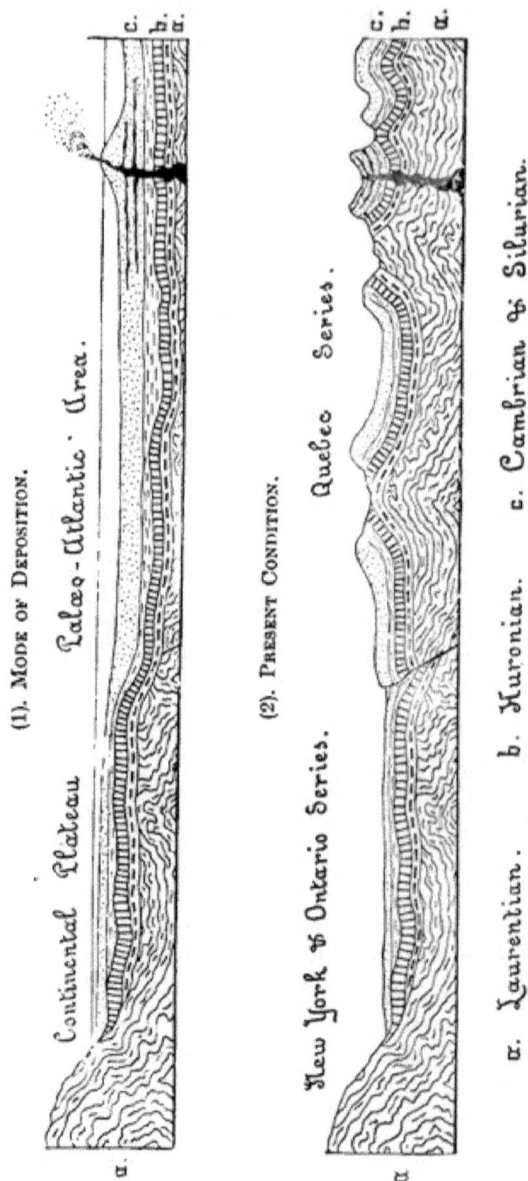

(1). MODE OF DEPOSITION.

Continental Plateau   Palæo-Atlantic · Area.

(2). PRESENT CONDITION.

New York & Ontario Series.   Quebec Series.

a. Laurentian.   b. Huronian.   c. Cambrian & Silurian.

Fig. 172.—Ideal Sections illustrating the manner of deposition of the Quebec Series of Logan, and its present relations to the older Palæozoic rocks of the central plateau of America.

represented in the Section, Fig. 172 (1.) Toward the close of the Siluro-Cambrian period there followed on the previous unequal elevation and depression, one of the great crumplings of the crust of the earth, which crushed the thick and then soft deposits of the Atlantic border against the edge of the inland Laurentian area, producing foldings, flexures, lateral thrusts and reversed faults in these marginal deposits, but leaving the thinner beds on the Laurentian plateau comparatively undisturbed. Thus was produced the state of things represented in Fig. 172 (2), which shows the present condition of the Quebec group strata. The beds of the Quebec Series which represent a portion of the Atlantic deposits of the period, formed in comparatively deep and cold water, and containing remains of a fauna suited to such conditions, were thrust against the Laurentian plateau and the flat beds of the inland Potsdam, Calciferous and Chazy, and piled up in gigantic earth-waves. Lastly, the subsequent denudation has sculptured these folds and has probably, in some denuded anticlinals, exposed portions of older Cambrian and Huronian rocks. This is also shown in the section.

With these typical Quebec Group rocks, which are characterised by abundant remains of Trilobites, Graptolites and Sponges, there are associated altered rocks originally classified by Logan as metamorphosed members of the Quebec Group, but now regarded by Dr. Selwyn as Cambrian and Pre-Cambrian. The latter are thus described by him :—*

"The Palæozoic rocks lie unconformably on or against an axis of Pre-Cambrian sub-crystalline rocks, hydro-mica slates, quartzites, crystalline dolomites, diabase, gabbro, olivinite, serpentine and volcanic agglomerates, the lowest beds of the axis being micaceous and granitoid gneisses. Many of these volcanic agglomerates and diabases are now serpentines. This lower portion of the Quebec group is defined on the geological map and colored as Pre-Cambrian. It extends from Sutton Mountain, on the Vermont boundary, to a point some miles north of the latitude of Quebec city, where it becomes covered by the unconformable junction of the Levis formation with the Siluro-Devonian rocks of the Gaspé series. To the north-east it again appears in some of the prominent peaks and ridges of the Shickshock Mountains, the northern flanks of which are occupied by the Cambrian and Cambro-Silurian formations of the Quebec Group, and the southern by those of the Siluro-Devonian system above referred to, as being in contact with the former a short distance to the south-west of the mountains."

"Similar Pre-Cambrian rocks form also several subordinate ridges to the south-east, and as in the main axis, they are everywhere characterized

* Descriptive sketch of Geology of Canada, 1884.

by the presence of sulphuretted copper ores; also hematites, magnetite, chromic iron and ores of antimony. The magnetite is, for the most part, economically unavailable on account of the high percentage of titanic acid. The soap-stone, pot-stone or mica-rock, serpentine and asbestos, described in the Geology of Canada, also belong to the Pre-Cambrian belts. One of these belts crosses the St. Francis River between Sherbrooke and Lennoxville. It constitutes the high ridges known as the Stoke Mountains, between Lake Massawippi and Little Magog, and in it are the most extensively worked copper mines of Canada. No fossils of any kind have yet been found in the rocks of these belts, and they are presumed to belong to the Huronian System, not only because of the geological position which they apparently occupy, but also on account of their close correspondence with it in physical aspect, and in mineral and lithological characters."

In this region are granites of Devonian age, trappean masses of Silurian age, and probably still older igneous rocks, some of them associated with large masses of serpentine, and the Siluro-Cambrian rocks are in places altered into hydro-mica schists and graphitic slates.

In this district are the valuable copper mines of the Eastern Townships, the gold deposits of the Chaudière, &c., and the deposits of Chrysotile or fibrous Serpentine usually named Asbestos and now extensively worked; also quarries of granite and marble.

3. The third and fourth districts are characterised by the prevalence of undisturbed members of the Siluro-Cambrian, Silurian and Erian systems. The first mentioned system prevails in the Province of Quebec where its ordinarily flat surface is broken by abrupt eminences of igneous rock ejected at the close of the Siluro-Cambrian, and affording excellent opportunity for the study of intrusive rocks and the remains of ancient volcanic vents, in the hills of Rigaud, Montreal, Montarville, Beloeil and Monnoir, Rougemont, Yamaska, Brome and Shefford. These hills form a line of ancient volcanoes extending across the Siluro-Cambrian plain for a distance of 90 miles.

The Siluro-Cambrian occupies the north side of Lake Ontario, but at the head of the lake it is covered by the Silurian, the thick limestones of which form the great Niagara escarpment, which is so prominent a geographical feature in the Lake region, and gives origin to the fall of Niagara. In Western Ontario, between Lakes Huron and Ontario, the Silurian is overlaid by the Erian or Devonian, of which in this region the Oriskany sandstone, the Corniferous limestone, remarkably rich in silicified corals, and the Hamilton shales, are characteristic features.

The Palæozoic rocks of these districts having been already noticed in Part II., it will be necessary here only to refer to the geographical distribution and most characteristic exposures.

The oldest Cambrian rocks are represented in the third district only by the Georgia series, which enters into Canada and occupies a limited space in the vicinity of Phillipsburg, and has been described by Mr. Billings.* The Potsdam sandstone in this district usually rests directly on the Laurentian, all the older groups being absent. It consists of siliceous sandstone often indurated into quartzite and showing ripple-marked surfaces. The prevalent fossils are the cylindrical perforations, usually tortuous and sometimes branching, known as *Scolithus Canadensis*, and on the surfaces of some of the beds are the crustacean footprints named *Protichnites*. Beds of conglomerate occur locally. The Potsdam covers considerable portions of Huntingdon, Chateauguay, Beauharnois, Soulanges and Vaudreuil, and skirts the Laurentian on the Lower Ottawa and in places as far east as the St. Maurice River. In the vicinity of Montreal it is well seen at St. Anne's, Vaudreuil, Beauharnois, and at various points on the Ottawa River.

The Calciferous, a formation consisting largely of dark-coloured impure dolomite, immediately overlies the Potsdam and in places graduates into it. It covers considerable areas adjoining the Potsdam in Huntingdon, etc., and is well seen at St. Anne's, near Montreal, near Beauharnois and Lachute. In these places it affords several characteristic fossils, as *Ophileta compacta, Murchisonia Anna, Piloceras amplum, Orthoceras,* etc. The Calciferous is regarded as a contemporaneous formation with the older parts of the Quebec Group, and notably with the Dictyonema beds of Matane and Cape Rosier. Rocks of this age also skirt the Laurentian on the north shore of the Gulf of St. Lawrence at Mingan and elsewhere.

The limestones and shales of the Siluro-Cambrian series occupy the greater part of the remainder of the Lower St. Lawrence plain, forming a broad belt on both sides of the St. Lawrence, and also occupy the country between the Lower Ottawa and the former river. This system includes all the formations from the Chazy to the Hudson River, inclusive, and in the district in question the Trenton and Utica formations occupy the widest areas, with patches of Hudson River superimposed, and a margin of Black River and Chazy limestones between them and the older rocks. These formations are well seen in the quarries and other excavations in the vicinity of Montreal and Ottawa.

---

* Geology of Canada, p. 283, et. seq. See also Walcott, Repts. of U.S. Geol. Survey.

A small patch of Silurian limestone of Helderberg age occurs in the Island of St. Helen, at Montreal; and there are good exposures of Silurian rocks, some of them rich in characteristic fossils, on the north side of the Baie de Chaleurs, and on the south side near Dalhousie, and also in the Peninsula of Gaspé. There are also considerable areas of Silurian rocks, in some places altered and with slaty structure, in the hilly country extending southwardly from Gaspé, on the confines of Quebec and New-Brunswick.

Anticosti exhibits on its south side a large area of limestones of the lower part of the Silurian, rich in characteristic fossils, while the northern part of the island has a belt of Hudson River rocks, and these seem to pass upward gradually into the Silurian.

The Erian or Devonian series is represented in this district only by the Gaspé sandstones, rich in fossil plants, while the lower beds of the Carboniferous occupy a limited area on the north side of the Baie des Chaleurs. No geological system later than the Carboniferous has been recognised in the district except the Pleistocene.

The Pleistocene deposits are spread over all the lower parts of the Provinces of Quebec and Ontario, and consist of marine beds in the valley of the St. Lawrence, with evidences of local glaciers on the hills. The prevailing direction of drift is from the north-east, and on the hills south of the Lower St. Lawrence there is evidence of movement of material from the high grounds both to the north and south.*

In the St. Lawrence Valley these deposits may be tabulated as follows, in ascending order:—

(a) Lower stratified sands and gravels (Syrtensian deposits of Matthew).

These represent land surfaces and sea and coast areas immediately anterior to the Boulder-clay period.

(b) Boulder-clay or Till; hard clay, or unstratified sand, with boulders, local and travelled, and stones often striated and polished. It rests on striated surfaces.

The Lower St. Lawrence region holds a few marine shells of Arctic species. Farther inland it is non-fossiliferous, but has usually the chemical characters of a marine deposit.

(c) Lower Leda clay; fine clay, often laminated, and with a few large travelled boulders, probably equivalent to Erie clay of inland districts.

Holds Leda (Portlandia) arctica, and sometimes Tellina groenlandica; and seems to have been deposited in very cold and ice-laden water.

(d) Upper Leda clay, and probably Sangreen clay of inland districts; clay and sandy clay, in the Lower St. Lawrence, with numerous marine shells.

Holds in Eastern Canada a marine fauna identical with that of the northern part of the Gulf of St. Lawrence at present: and locally affords remains of boreal flora.

* Chalmers, Trans. R.S.C. The Author in Canadian Record of Science.

(e) Saxicava sand and gravel, often with numerous travelled boulders (Upper Boulder deposit), probably the same with Algoma sand, etc., of the West. — Shallow-water fauna of boreal character, more especially *Saxicava rugosa* and its varieties. Bones of Whales, etc.

(f) Post-Glacial deposits, river alluvia and gravels, Peaty deposits, Lake bottoms, etc. — Remains of *Mastodon* and *Elephas*, modern fresh-water shells.

All of these deposits are seen in the vicinity of Montreal.

The lower Boulder-clay (*c*) is often a true and very hard Till, resting on intensely glaciated rock-surfaces, and filled with stones and boulders. Where very thick, it can be seen to have a rude stratification. Even when destitute of marine fossils, it shows its submarine accumulation by the unoxidized and unweathered condition of its materials. The striæ beneath it, and the direction of transport of its boulders, show a general movement from N.E. to S.W., or up the St. Lawrences Valley from the Atlantic. Connected with it, and apparently of the same age, are evidences of local glaciers descending into the valley from the Laurentian highlands.

The boulder clay and superficial deposits of the basin of the great lakes are of similar character, but they are destitute of marine shells, contain land plants in the beds corresponding to the Leda clay, and are surmounted in places by old lake margins evidencing a former greater extension of the lakes.

In the Ontario Region the Pleistocene and modern formations, which may for comparison be noticed here, have been divided as follows :—

    (*b*) Boulder clay.

    (*c*) Erie clay.

    (*d*) Saugeen clay and sand.

    (*e*) Artemesia gravel and Algoma sand.

    (*f*) Recent alluvia,

which correspond in a general way to the beds designated by the same letters in the preceding table.

The Lower Leda clay (*d*) seems in all respects similar to the deposits now forming under the ice in Baffin's Bay and the Spitzbergen Sea. The Upper Leda Clay represents a considerable amelioration of climate, its fauna being so similar to that of the Gulf of St. Lawrence at present that I have dredged in a living state nearly all the species it contains, off the coasts on which it occurs. Land plants found in the beds holding these marine shells are of species still living on the north shore of the St. Lawrence, and show that there were in certain portions of this period considerable land surfaces clothed with vegetation. The Upper Leda Clay is probably contemporaneous with the so-called inter-glacial deposits holding plants and insects discovered by Hinde on the shores of Lake Ontario.*   On the Ottawa it contains land plants of modern Canadian species, insects and feathers of birds, intermixed with skeletons of capelin and shells living in the Gulf of St. Lawrence.

The changes of level in the course of the deposition of the Leda clays must have been very great ; fossiliferous marine deposits of this age being found at a height of at least 600 feet, and sea-beaches at a much greater elevation, while at other times there must have been large land areas and even fresh-water lakes.   Littoral gravels and sands of this period may also be undistinguishable, except by their greater elevation, from those of the Saxicava sand.   I have recently described the bones of a large whale (*Megaptera longimana*) from gravel north of the outlet of Lake Ontario and 420 feet above the level of the sea, which is not improbably contemporaneous with the Leda clay of lower levels, and much higher than deposits near Lake Ontario regarded as of lacustrine origin.†   These changes of the relative levels of sea and land must be taken into account in explaining the distribution of marine clays and sands, boulder deposits, etc., which are often studied merely with reference to the present levels of the country, or as contemporaneous deposits without regard to their elevation, a method certain to lead to inaccurate conclusions.

The Saxicava sand (*f*) indicates shallow-water conditions with much driftage of boulders, and probably glaciers on the mountains.   It constitutes in many districts a second boulder formation, and possibly implies a somewhat more severe or at least more extreme climate than that of the Upper Leda clay.   Terraces along the coast mark the successive stages of elevation of the land in and after this period.   There is also evidence of a greater elevation of the land succeeding the time of the Saxicava sand, and preceding the modern era.‡

It is well known that very diverse theoretical views exist among geologists as to the origin of the deposits above referred to.   The conclusions

---

* Proceedings of Canadian Institute, 1877.   Dr. Hinde in this paper states that the Leda clay belongs to the " close of the Glacial Period," and that boulder-drift is not found above it.   In truth, however, as Admiral Bayfield, Sir Charles Lyell, and the writer have shown, boulder-drift is still in progress in the Gulf and River St. Lawrence, though in a more limited area than in the Pleistocene period ; but any considerable subsidence of the land might enable it to resume its former extension.

† Canadian Naturalist, vol. x. No. 7.

‡ Supplement to Acadian Geology, 3rd edition, pp. 14, et. seq.

which have been forced upon the writer by detailed studies extending over the last forty years, are that in Canada the condition of extreme glaciation was one of partial submergence, in which the valleys were occupied by a sea laden with heavy field ice continuing throughout the summer, while the hills remaining above water were occupied with glaciers, and that these conditions varied in their distribution with the varying levels of the land, giving rise to great local diversities, as well as to changes of climate. There seems to be within the limits of Canada no good evidence of a general covering of the land with a thick mantle of ice, though there must at certain periods have been very extensive glaciers on the Laurentian axis and in the mountainous regions of the west. It does not, indeed, seem possible that, under any conceivable meteorological conditions, an area so extensive as that of Canada, if existing as a land surface, should receive, except on its oceanic margins, a sufficient amount of precipitation to produce a continental glacier.

Details on some of the above mentioned formations will be found in my "Notes on the Post-Pliocene of Canada," and a large amount of recent information exists in the Reports of the Geological Survey of Canada, and in papers published in the Canadian Naturalist and Geologist, and in the Canadian Record of Science.

4. The Western Palæozoic district extends to the south-west of the Thousand Islands in the St. Lawrence valley, and includes the north shore of Lake Ontario, between Lakes Erie and Huron, and the Peninsula extending to Cabot's Head and the Great Manitoulin and its associated islands.

In this district the Niagara limestone forms a well marked escarpment, extending from Niagara Falls westward and northward to Cabot's Head and Manitoulin Island. On the east of this the country north of Lake Ontario is occupied by the formations from the Hudson River to the Potsdam, inclusive, while to the west the peninsula between Erie and Huron is occupied with the upper members of the Silurian and the Erian.

All the formations of this district are of similar type to those of the interior plateau region of North America, commonly known as the New York type of the Palæozoic formations.

The Potsdam sandstone is not very extensively distributed, except in the eastern part, but the exposures of the upper layers at Beverley is remarkable for its fine specimens of *Lingula acuminata*, and a quarry near Perth has afforded some of the finest specimens of the tracks known as *Protichnites* and *Climactichnites*. In a great part of this district the older beds seem to have been overlapped by the extension of the Siluro-Cambrian limestones, whose outcrop crosses in a broad band from Kingston to Georgian Bay and has on its south-west side a wide area of Utica and Hudson River beds. The Trilobites *Asaphus Canadensis* and species of *Triarthrus* seem to be specially characteristic of this district, and the former occurs abundantly at Collingwood as well as near to Ottawa, where, however, the Trenton limestone is well exposed and abounds in characteristic fossils, especially in crinoids. At various places near Toronto the Hudson River beds are well exposed. On the Don and Humber, *Leptaena sericea, Modiolopsis modiolaris, Ambonychia radiata, Pterinea demissa* and *Orthoceras crebriseptum* are characteristic species. Farther west, in Manitoulin Island, the corals *Petraia corniculum* and *Faristella stellata* are very abundant.

At the western end of Lake Ontario the Medina sandstone, constituting the base of the Silurian, spreads somewhat widely on the lake shore, and toward the foot of the escarpment of Hamilton, where it presents a thickness of about 600 feet of red and gray sandstone and marl. In its westward extension to Georgian Bay it thins off and disappears. Two characteristic fossils are the tracks *Arthrichnites Harlani* and the brachiopod *Lingula cuneata*, an Orthoceras (*O. multiseptum*) occurs rarely and a few lamellibranchiate shells (*Modiolopsis.*)

The great escarpment west of Hamilton and extending to Cabot's Head and by which the falls of Niagara are produced, exhibits the Clinton group and the Niagara shale and limestone. At the falls of Niagara thick beds of Dolomite and dolomitic limestone constitute the most prominent features ; and the most abundant fossils are *Stromatoporae*, but below the falls and near Hamilton as well as in many other localities along the escarpment, numerous fossils have been found. *Astylospongia* and other sponges, *Favosites Niagarensis* and other corals, *Pentamerus oblongus, Spirifer Niagarensis*, different species of *Dictyonema* and other graptolites, *Calymene Niagarensis* and *Dalmanites limulurus* are characteristic.

The Guelph formation is peculiar to this district, and is a dolomite containing a special fauna belonging apparently to a land-locked sea, and in which the bivalve *Megalomus Canadensis* and the curious Brachiopods of the genus *Trimerella* are characteristic. Good exposures occur near Guelph and at Elora, where many forms of Stromatoporae occur with other fossils.

The Salina or Onondaga salt group is well developed. Its outcrop extends from the Niagara River above the falls to Lake Huron, and it contains valuable deposits of rock salt and gypsum. Its structure has been noticed in the previous part and it undoubtedly marks a period of continental elevation and dessication, separating the Silurian period into two portions. (See section at Goderich in Part II., p. 80.) The Upper or Helderberg member of the Silurian is not largely developed in this district, and is especially characterized by the Merostomatous crustacean *Eurypterus remipes*.

The Oriskany sandstone constitutes the base of the Erian deposits, and is here as elsewhere rich in *Spirifer arenosus*, *Renssellaria ovalis* and several species of *Favosites*. Many of its species are identical with those of the Upper beds of the Silurian, but Erian fossils rather preponderate. It may, however, be regarded as a transition group between the Silurian and Erian. The outcrops of this and the following series occur in succession, crossing the peninsula between Lakes Ontario and Huron.

The Corniferous formation is one of the most remarkable in the district in a palæontological point of view. It is especially rich in silicified corals, which exhibit in perfection, unequalled elsewhere, the profusion of fossils of this kind in the Erian seas. Billings has described a large number of species of corals from this formation, which also abounds in Brachiopod shells. Among the corals are species of *Favosites*, *Michellena*, *Fistulipora*, *Zaphrentis*, *Cystiphyllum*, *Haimeophyllum*, *Phillipsastrea*, *Syringopora*, *Eridophyllum*, etc., and among the Brachiopods are species of *Strophomena*, *Chonetes*, *Leptocoelia*, *Streptorhynchus*, *Orthis*, *Rhynchonella*, *Pentamerus*, *Stricklandinia*, *Spirifer*, *Athyris*, *Charionella* and *Centronella*. Many other fossils of other types also occur.

The Hamilton shales are next in succession, and it is from borings in these shales, and especially along anticlinal bends of the strata, that the petroleum of the Province of Ontario is chiefly obtained. The Hamilton shales afford many fossils, which for the most part are identical with those of the Corniferous, but some forms, as *Spirifer mucronatus* and *Atrypa reticularis* are especially abundant and characteristic.

The upper members of the Erian, the Portage and Chemung groups of the United States, are only slenderly represented in Ontario. Kettle Point in Lake Huron is especially remarkable for beds of bituminous shale filled with the macrospores of Rhizocarps known as *Sporangites Huronensis*, and which not improbably indicate the character of the vegetation from which the petroleum of the Erian has been derived, as remains of these aquatic plants abound throughout the Erian formations.

No later formations occur in this district except the Pleistocene and Modern, which have been already mentioned. Spencer has ascertained

that many of the peculiarities of the great lakes may be accounted for
by warping, or unequal elevation of the land in and since the Pleistocene,
and it is now known that these great sheets of fresh water have been
dammed up partly by unequal elevation but more by filling of their old
outlets with Pleistocene deposits, so that they are very modern features,
and their basins must have been excavated before the Pleistocene, and probably at a time when the land was more elevated than at present.

Bones of the *Mastodon* and of a fossil Elephant—*Euelephas Jacksoni*
—have been found in Post-glacial gravel and peat in the vicinity of
Lake Ontario.

Conspectus of Geological Formations in the Canadian Region, with Typical Localities, described in Reports of the Geological Survey.—

Quaternary, Saxicava Sand ⎫ Vicinity of Montreal, Beauport, near
    Leda Clay.   ⎬ Quebec. R. du Loup. (In Ontario
    Boulder Clay or Till. ⎭ Erie Clay and Saugeen sand.

Lower Carbonif., Bonaventure, Baie des Chaleurs.

Devonian, Catskill.   * Gaspé Bay, Scaumenac Bay.

 "   Chemung and Portage. Kettle Point, Lake Huron.

 "   Hamilton, including⎫
    Marcellus and Ge-⎬Western Ontario, Widder, etc.
    nesee.   ⎭

 "   Corniferous or Upper⎫
    Helderberg.   ⎬N. Shore of Lake Erie.
 "   Oriskany.   ⎭

Silurian, Lower Helderberg. St. Helen's Island, near Montreal.
 "   Salina or Onondaga. Goderich, Ontario.
 "   Guelph.   Guelph and Galt, Ontario.
 "   Niagara.   Niagara Falls, vicinity of Hamilton.
 "   Clinton.   Manitoulin Island, Niagara, etc.
 "   Medina and Oneida. Hamilton, etc.

Siluro Cambrian, Hudson R. Chambly, Toronto, etc.

 "   Utica.   ⎧Near Montreal and Ottawa ; Whitby,
        ⎩ Ontario ; Collingwood, Ontario.

        ⎧Near Montreal, Point Claire, Caugh-
 "   Trenton. ⎪ nawaga, Belleville, Kingston, Ottawa.
 "   Chazy.   ⎨ (Quebec group—Levis and S. Shore
        ⎩ L. St. Lawrence.)

Cambrian, Calciferous.   ⎧St. Anne, Beauharnois.
        ⎩Matane, Cape Rosier, etc.

Lower Cambrian, Potsdam. St. Anne, Perth, etc.
 "   " Georgia. Phillipsburg.
 "   " Keweenian. Maimanse, Lake Superior.
 "   " Animikie. Port Arthur, and vicinity.

Eozoic or Archæan, Huronian. N. Shore, Georgian Bay.
 "   U. Laurentian. St. Jerome, Bay St. Paul.

        ⎧Lachute, Grenville, Calumet, Petite
 "   M. & L. Laurentian.⎨ Nation, Buckingham, N. Shore Lake
        ⎩ Superior.

III. THE REGION OF MANITOBA AND THE NORTH-WEST TERRITORIES.

The following account of this region is derived almost wholly from papers and reports of Dr. G. M. Dawson, F.G.S., etc.—

PHYSICAL GEOGRAPHY.

The northern part of the North American continent is geologically, and to a great extent also physically, divisible into two great portions. In the first, extending from the Atlantic coasts to the south-eastern edge of the Laurentian axis,—which is marked by a chain of great lakes stretching from the Lake of the Woods to the Arctic Ocean,—the Archæan plateau is the dominant feature, the succeeding formations arranging themselves about its edges or overlapping it to a greater or less extent in the form of bays or inlets; but, with the single exception of limited tracts of Triassic rocks, no mesozoic or tertiary strata are represented in it. In the second, stretching westward to the shores of the Pacific, the Archæan rocks play a very subordinate part, and Mesozoic and Tertiary rocks are abundantly represented and alone characterize the whole area of the great plains. Correlated with the difference of age in the formations represented, is the fact that at a date when the flexure and disturbance of the eastern region had practically closed, and it was set and firm, the western Cordillera belt continued to be the theatre of uplift and folding on a gigantic scale.

The great region of plain and prairie which occupies the central part of the continent is on the 49th parallel of north latitude, included in longitude between the 96th and 114th meridians. It narrows pretty rapidly northwards, by the encroachment on it of its eastern border, but continues as a great physical feature even to the shore of the Arctic Ocean, where it appears to have a breadth of between 300 and 400 miles. Beyond the North Saskatchewan River, however, it loses its essentially prairie character, and with the increasing moisture of climate, becomes, with limited exceptions, thickly covered with coniferous forest.

The north-eastern boundary of this interior continental plateau, north of latitude forty-nine, is formed, as above stated, by the south-western slope of that old crystalline nucleus of the continent which extends north of the St. Lawrence and Great Lakes from Labrador to the Lake of the Woods, with a general east and west course, and then, turning suddenly at an angle of 60° to its former general direction, runs with a north-north-west course to the Arctic Sea. The eastern barrier is rather a rocky plateau than a mountain region. It presents no well-defined height of land, and the watershed-line follows a very sinuous course among the countless lakes, small and great, which cover its surface. Northward from the Lake of the Woods, it divides the waters flowing into Hudson's Bay from those draining directly into the Arctic Ocean, with one important exception. The Nelson River, carrying the accumulated waters of the Saskatchewan, the Red River and innumerable smaller streams, breaks through the Laurentian plateau at the north end of Lake Winnipeg, and empties into Hudson's Bay at York Factory. The Churchill or English River, a not inconsiderable stream, passes through the same gap.

Near the 49th parallel, the Rocky Mountains on the west rise abruptly from the elevated plain at their base, and often present to the east almost perpendicular walls of rock. A short distance farther north, however, they become bordered by an important zone of foot-hills composed of crumpled Mesozoic rocks, and these continue with varying breadth at least as far north as the Peace River region. Between the fifty-first and fifty-second parallels the Rocky Mountain range appears to culminate, and to the north gradually decreases in elevation till on the borders of the Arctic Ocean it is represented by comparatively low hills only. With this decrease in height the mountains become a less complete barrier, and the streams flowing eastward across the plains rise further back, till in the cases of the Peace and Liard Rivers the waters from the central plateau of British Columbia completely traverse the range.

The whole interior region of the continent slopes gradually eastward from the elevated plains lying near the base of the Rocky Mountains to the foot of the Laurentian highlands, and though the inclination is more abrupt in approaching the mountains, it is not so much so as to attract special attention. Between the fifty-fourth and forty-ninth degrees of latitude, however, along two lines which are in a general way parallel and hold a north-west and south-east course across the plains, more or less definite step-like rises occur. These respectively form the eastern boundaries of the two higher prairie plateaus, and the most eastern of them overlooks the lowest prairie level or that of the Red River valley. The three areas of plains or "prairie steppes" thus outlined differ considerably in age and character, and their margins have been impressed on the soft formations of the plains by the action of sub-aerial denudation, by that of former lakes and probably also by that of the sea. Though not everywhere sharply defined in nature, they may be considered separately for purposes of description.

The actual increase of elevation accounted for in the two escarpments, however, is slight compared with that due to the uniform eastward slope of the plains. The direction of greatest inclination is toward the northeast, and a line drawn from the intersection of the forty-ninth parallel and the mountains, to a point on the first prairie-level north of Lake Winnipeg, will be found to cross the escarpments nearly at right angles, and to have an average slope of 5.38 feet to the mile. From the same initial point, in a due east line to the lowest part of the valley of the Red River—a distance of 750 miles—the plains have an average slope of 4.48 feet per mile.

The first or lowest prairie-level is that of which the southern part lies along Red River, and which northward embraces Lake Winnipeg and associated lakes and the flat land surrounding them. A great part of its eastern border is conterminous with that of Lake Winnipeg, and formed by the rocky front of the Laurentian, while east of Red River it is bounded by the high-lying drift terraces surrounding the Lake of the Woods and forming part of the drift plateau of northern Minnesota. To the west it is limited by a more or less abrupt edge of the second prairie level, forming an escarpment, which, though very irregular in some places, is scarcely perceptible where the broad valley of the Assineboine breaks through it. The escarpment, where it crosses the forty-ninth parallel, is known as Pembina mountain, and is continued northward by the Riding, Duck, Porcupine and Basquia Hills. The average height above the sea of this lowest level of the interior continental region is about 800 feet, the lowest part being that including the Winnipeg group of lakes which have an elevation of about 700 feet. From this it slopes up southward, and attains its greatest elevation—960 feet—at its termination about 200 miles south of the international boundary. The edges of this plain are also, notwithstanding its apparent horizontality, considerably more elevated than its axis where this is occupied by the Red River. Its width on the forty-ninth parallel is fifty-two miles only. Its area north of the same line may be estimated at 55,000 square miles, of which the great system of lakes in its northern part occupies about 13,900 square miles. A great part of this prairie-level is wooded more or less densely, particularly that portion adjacent to the lakes. The southern part, extending southward from Lake Winnipeg, includes the prairie of the Red River valley with an area north of the forty-ninth parallel of about 6,900 square miles.

The superficial deposits of this area are chiefly those of a former great lake, which has been named by Mr. Warren Upham Lake Agassiz, and which occupied it toward the close of the glacial period. Many of the old shore lines and beaches of this lake have been traced out, and it has been shown by General Warren, Dr. G. M. Dawson, and the first named gentleman, that its outflow was originally southward to the Mississippi. The fine silty material now flooring the Red River plain and constituting its soil of unsurpassed fertility was laid down in this ancient lake. The Red and Assiniboine Rivers have not cut very deeply into these superficial deposits, having already nearly reached a base level of erosion, and the surface of the plain is level and little furrowed by denudation.

The second steppe of the plains is bounded to the west by the Missouri Coteau and its northern continuations constituting the edge of the third steppe. Its width on the forty-ninth parallel is two hundred and fifty miles, and on the fifty-fourth about two hundred miles, though it cannot there be so strictly defined. Its total area between these two parallels is about 105,000 square miles, and includes the whole eastern portion of the great plains, properly so called, with an approximate area of 71,000 square miles. These occupy its southern and western portion, and are continuous westward with those of the third steppe. The present rivers have acted on this region for a much longer time than on the last, and with the advantage of a greater height above base level ; and now flow with uniform though often swift currents, in wide trough-like valleys excavated in the soft material of the plains, and frequently depressed from one hundred to three hundred feet below the general surface. In these valleys the comparatively insignificant streams wander from side to side in tortuous channels, which they leave only at times of flood. The surface of this steppe is also more diversified than the last, being broken into gentle swells and undulations, partly due to the present denuding agencies, but in part also to original inequalities in the deposition of the drift material which constitutes the superficial formation. The average altitude of this region may be stated as 1600 feet, and its soil and adaptability for agriculture differ considerably in its different portions, though it is generally fertile.

The third or highest steppe of the plains may be said to have a general normal altitude of about 3000 feet, though its eastern edge is usually little over 2000 feet, and it attains an elevation of 4000 feet at the foot of the Rocky Mountains. Its area between the parallels above defined, and including the high land and foot-hills along the base of the mountains, is about 134,000 square miles, and of this the greater part is almost entirely devoid of forest, the wooded region being chiefly confined to a portion of its northern and north-western extension near the North Saskatchewan River and its tributaries. Its breadth on the forty-ninth parallel is four hundred and sixty-five miles, and its eastern boundary is there well marked, being the broken hilly country known as the *Côteau de Missouri*, or Great Coteau. This crosses the International boundary near the one hundred and fourth meridian, and thence runs east of the Old Wives' Lakes to the South Saskatchewan, and is then continued to the north by a range of high lands, of which the Eagle Hills constitute part, to the elbow of the North Saskatchewan, and beyond that river probably to the Thickwood Hills.

This portion of the great plains is much more diversified than either of those before described. It has been elevated to a greater height above the sea level, and acted on to a much greater extent by eroding agents, both in later Tertiary time and subsequent to the glacial period. Those portions of its surface which still remain but little modified, form table-lands such as those of the Cypress Hills and Wood Mountain. The immense denudation which has taken place is evidenced by the size and depth of the valleys of the streams, the great ravines and "coulées" which have been cut and are still extending themselves among the soft sandstones and clays of the Cretaceous and Laramie formations, and the isolated plateaus and buttes which now stand far out on plains of lower level, seamed with newer systems of drainage. Deposits belonging to the glacial period, with transported boulders and gravel, are found over almost the entire area of the highest steppe, but are spread less uniformly than on the lower levels, and the surface is often based almost directly on the Cretaceous and Laramie beds. This is ample proof that previous to the glacial period the surface was much more rugged and worn than it is, now that the glacial deposits have filled many of the deeper hollows and given rounded and flowing outlines to the whole.

In the foot-hills of the Rocky Mountains the previously undisturbed beds of the plains are thrown into wave-like flexures and compressed folds, which the surface participates in to a lesser degree, assuming the form of crest-like parallel ridges which frequently possess considerable uniformity. The nature of the soil and prospective agricultural value of this great district are too varied to allow of generalization. Though it must be regarded rather as a grazing than a farming region, it presents frequently an excellent soil, and when the rainfall is sufficient and the altitude not too great, considerable connected tracts may yet be brought under cultivation.

North of the North Saskatchewan no extensive treeless plains occur in the central region of the continent, and the forest country of the east forms a wide unbroken connection with that of the northern part of British Columbia, and though prairies of very attractive character are found near the Peace River, they are limited in area and isolated by belts of woodland. The width of the Mesozoic and Tertiary plain gradually diminishes to the north, being less than 400 miles near the fifty-sixth parallel, and it is possibly completely interrupted north of the sixty-second parallel by the inosculation of the palæozoic rocks of the east and west. In the basin of the Peace, the lower areas are covered superficially by fine silty deposits resembling those of the Red River valley, and doubtless indicating a former great lake or extension of the sea in the time immediately succeeding the glacial period.

Though thus so remarkably simple and definite in its grand features the interior region of the continent shows many irregularities in detail. The second steppe has some elevations on its surface as high as the edge of the third plateau, and that part surrounding the Assiniboine River and its tributaries is abnormally depressed, causing some portions of the eastern edge of this prairie-level which overlook Manitoba Lake, more to resemble outliers than integral parts of it.

The transverse water-sheds which bound the drainage area of the Saskatchewan and Red Rivers to the south and north, though comparatively low and diffuse, and insignificant as geological boundaries, are important geographically. Taken as a whole, however, the central portion of the Dominion may be regarded as a great shallow trough, of which the western edge is formed by the Rocky Mountains, the eastern by the Laurentian axis, but in which the western portion of the floor is now, (probably as the result of Post-Tertiary elevation,) higher than its eastern rim. Of the area as at first defined, extending from the fifty-fourth to the forty-ninth parallels, the great Saskatchewan River and its tributaries drain by far the largest part, or about 139,000 square miles. The Red River and its tributary the Assiniboine drain 70,500 square miles, and the valleys of the numerous small streams flowing into the Winnipeg group of lakes, including the area of the lakes themselves, drain 52,800 square miles. The upper branches of the Missouri, and especially those of its tributary the Milk River, drain a considerable area to the south, embracing about 22,800 square miles, while to the south of the first named parallel the tributaries of the Mackenzie drain an area of about 10,000 miles only. The total area of prairie country between the same limits, including that of all three steppes, may be estimated at

192,000 square miles. Though much of this area is not absolutely treeless the aggregate tree-clad area is relatively insignificant.

GEOLOGY.

The main geological features of the central and western portion of the Dominion have been touched on in connection with its physical structure. In the present section they are outlined in more systematic form.

The eastern margin of the great interior continental basin is composed of Silurian and Devonian rocks, which, resting almost horizontally on the upturned and greatly denuded margin of the Laurentian and Huronian, form a belt of varying width which appears to extend with little or no interruption from Minnesota to the shores of the Arctic Sea.

Recent investigations of the fauna of these rocks by Mr. Whiteaves show that the Trenton formation is most widely spread about the shores of Lake Winnipeg, while the horizon of the Hudson River is represented on the lake at the mouth of the Little Saskatchewan as well as at Stony Mountain in the Red River valley. In the vicinity of Selkirk in the same valley the fossils are those of the Galena limestone of the west, equivalent to the Utica. On Manitoba and Winnipegosis Lakes Devonian limestones occur.

The Siluro-Cambrian and Devonian rocks of the Red River and Winnipeg Lake region are for the most part pale grey or buff-coloured magnesian limestones. From Methy Portage northward it seems that Devonian rocks constitute almost the entire width of the Palæozoic belt. They appear on the Clearwater and Athabasca Rivers as bituminous limestones and shales, which are referred by Meek, from their fossils, to the Hamilton and Genesee formations. In this region these rocks yield large quantities of petroleum, which, exuding from them, saturates the overlying superficial materials and Cretaceous sandstones and gives rise to "tar springs" along the banks of the rivers. Salt springs also occur, and these are found to characterize the Devonian rocks southward to Red River, though no certain indications of petroleum have yet been observed south of Methy Portage.

The Cretaceous and Tertiary rocks occupy the entire breadth of the interior continental region from the belt of Siluro-Cambrian and Devonian limestones last noticed, to the eastern edge of the Cordillera belt. These rocks correspond in their widespread and homogeneous general character, to the uniformity of the great plains which they underlie, and may be characterized in a few words.

North of the forty-ninth parallel, the Systems represented—so far as at present known—are the Cretaceous and Tertiary. In the eastern portion of the region, the Cretaceous, owing to the mantle of glacial deposits, is in general, poorly exposed, but in most respects resembles the strata of the same age, studied by Messrs. Meek and Hayden in the corresponding portion of the Western States. It may be arranged as follows, in descending order :—

> Fox Hill Sandstones.
> Fort Pierre Clays.
> Niobrara Marls.
> Fort Benton Clays and Limestone.
> Dakota Sandstones and Clay.

For details see Pt. ii., p. 107.

So far as at present known, beds of the Dakota or lowest subdivision of the Missouri Cretaceous occupy a very small portion of the surface in Manitoba and the North-west Territory. Sandstones and shales possibly of this age occur at the base of the section on Swan River, west of Winnipegosis Lake, and fossil plants of this stage are found among the upturned bases of the Rocky Mountain foot-hills. The Benton and Niobrara sub-divisions (locally scarcely separable) are represented in the escarpment west of Winnipegosis and Manitoba Lakes by practically horizontal, dark gray clay shales interbedded with layers of chalky limestone and bands of sandstone. Similar rocks run southward along the base of the Pembina escarpment, the greater part of which is composed of Pierre shales. The lower dark shales met with in the vicinity of the Upper Milk River are also regarded as equivalent to the Benton, or Benton and Niobrara, while the Fort St. John shales of Peace River are referred to the Benton. It is probable that the Belly River and Dunvegan series, subsequently noticed, approximately represent the Niobrara, though differing much from more eastern developments of that sub-division.

The Pierre or next overlying group is in point of extent the most important of the sub-divisions of the Cretaceous in the North-West, its characteristic dark shales or shaly-clays underlying a great part of the prairie country. The shales frequently contain ironstone nodules, and in some places arenaceous layers also appear and are generally found to become more important in approaching the mountains. The lithological character of this group is, however, on the whole, remarkably uniform, and beds are occasionally found in it which are highly fossiliferous. The highest beds of the Cretaceous system proper, the Fox Hill, are closely related to the Pierre, and form in many places a series of passage beds between the Cretaceous and the overlying Laramie formation. When most characteristically developed they consist of sandstone and yellowish sandy shales.

Notices of the fossils of the western Cretaceous will be found in Part II., and more full deiails as to the marine invertebrates in the Reports on Mesozoic Fossils by Mr. Whiteaves of the Geological Survey.

Overlying the Cretaceous proper in perfect conformity, is a great series of estuarine and fresh-water beds which may collectively be referred to the Laramie formation. No question in western geology has given rise to so much discussion as that of the Cretaceous or Eocene age of these beds. It is one which depends almost entirely on the apparently conflicting evidence of the vertebrate, molluscan and vegetable fossils which they contain, and one which cannot here be entered into. It may suffice to note that while the vertebrates are of types regarded as Cretaceous, the plants and molluscs resemble most closely those of Tertiary beds of other regions. The molluscs being, however, for the most part, fresh-water forms, are not so characteristic as marine shells would be. The formation is really a transition one, but is closely attached to the Cretaceous by its complete stratigraphical conformity with that formation.

The most eastern locality of these beds is Turtle Mountain, on the forty-ninth parallel, where they form a considerable outlier. On the Souris River they are largely developed, and constitute the superficial formation of the whole country. The Laramie of this region is, however, an extension of that special development of the formation which on the Missouri has long been known as the Fort Union series. The rocks are generally soft, sandy clays and sandstones, of pale colours, and on the Souris hold ironstone and many seams of lignite of fair quality. Further west the typical Laramie covers a vast area and becomes a distinctly estuarine formation at the base. The western Laramie, particularly in the vicinity of the mountains, is largely composed of sandstones which are frequently hard, and the formation has a thickness of several thousand feet, and has been sub-divided into several groups.

An interesting area of Miocene Tertiary rocks occurs in the Cypress Hills, overlying both Cretaceous and Laramie beds. A small outlier of the same age caps the Hand Hills further north. The beds are chiefly pebble-beds, or conglomerates composed of coarse rolled shingle which has had its origin in the Rocky Mountains. These are associated with soft sandstones and sandy clays, and have yielded characteristic vertebrate remains, including those of two species of *Menodus (Brontotherium)*. The formation has a thickness of 500 feet or more in some places. It covers an area of 1,400 square miles in all.

It has already been stated that the Cretaceous rocks of the western part of the plains differ from those of the typical section first quoted. In the region of the Bow and Belly Rivers, and northward to the Saskatchewan, the Pierre is underlaid by an extensive fresh and brackish water series, consisting of sandy argillites and sandstones. This has been called the Belly River series, and appears to correspond precisely to that occupying a similar stratigraphical position on the Peace River, and there designated the Dunvegan series. It indicates the occurrence of a prolonged interval in the western Cretaceous area, during which the sea was more or less excluded from the region, and its place occupied for long periods by lagoons or fresh-water lakes.

Though the Dakota series is the lowest representative of the Cretaceous known to occur in the area of the plains, and has there been observed to rest on Palæozoic rocks, still older Cretaceous beds come to the surface in the corrugated region of the foot-hills, and exist also in more or less completely isolated troughs or in folds in the Rocky Mountains north of the forty-ninth parallel. These beds have a thickness of, at least, 7000 feet and have been named the Kootanie series. They contain a peculiar flora consisting largely of cycads and conifers, and at their base marine fossils of lower or middle Cretaceous age have been found. It is probable that the shore-line of the sea in which these beds were laid down occupied a position not far to the east of the present border of the foot-hills.

If we regard the Dakota, Benton and Niobrara as Middle Cretaceous (see Part II.), the Kootanie may be taken to represent a part, at least, of the Lower Cretaceous. This corresponds with the indications of the fossil plants; for while those of the Kootanie beds are cycads, conifers and ferns, akin to those of the Wealden and Triassic of other countries, those of the Dunvegan and Belly River are of more modern type like those of the Upper Cretaceous elsewhere.

The differences of opinion among geologists in the United States respecting the age of the Laramie may be partly explained by the development of this formation in Canada. In the Western or Bow River district this formation may be tabulated as follows, in descending order :—

1. *Porcupine Hill beds.*—Sandstones and clays with lignite and fossil plants.

2. *Willow Creek Series*—in which red clays predominate, and which has afforded no fossils.

3. *St. Mary River Series.*—Chiefly grey clays with fossil plants.*

---

* Reports of Geol. Survey for details.

In the Souris district, nearer the boundary of the United States, as stated in Part II., the upper member is represented by the continuation of the Fort Union group of the Missouri and the lower member by the Bad Lands group, the middle member being absent.   In both districts the upper member contains a rich flora of modern genera of exogens, such as *Populus, Platanus, Corylus,* &c., and the species resemble those of the Eocene elsewhere, and there are also Molluscs of Eocene type.*   The lower member holds a flora similar to that of the Belly River Series in the upper Cretaceous, and has also afforded reptilia remains of creta- ceous types.   It is probable, therefore, that the Laramie may ultimately be divided, the lower part being placed with the Cretaceous and the upper with the Eocene.   The Dinosaurs and small mammals described by Marsh, from the Laramie of the United States, and referred to the Cretaceous probably belong to the lower member.†

The Cretaceous and Laramie beds of the whole eastern portion of the interior Continental region are almost absolutely horizontal, or affected by such slight inclinations that no dip is observable in individual sections. The beds of both series have, however, participated in the western uplift of this part of the Continent, and are found at ever-increasing angles on approaching the Rocky Mountains.   Near the base of the range they are also found to show more pronounced undulations, and in a narrow belt along the foot of the mountains are sharply folded and contorted.   Iso- lated areas of these newer rocks have also been found in the Rocky Mountains themselves.

The most important study depending on the question of the Cretaceous and Laramie rocks of the North-West, is that of the fuel supply.   In the eastern region, lignites of fair quality and workable thickness occur in the Laramie rocks of the Souris district, but have so far not been found in the underlying Cretaceous.   Further west the Cretaceous also becomes a coal-bearing formation, and in the vicinity of the Bow and Belly, import- ant lignites or coals have now been found in the Kootanie series (anthra- cite of Bow Pass,) Belly River series (Medicine Hat, etc.); base of the Pierre (Lethbridge, etc.); top of the Pierre (Bow River) and in the lower subdivision of the Laramie (Blackfoot Crossing, etc.)   In the Peace River district seams which may prove to be of a workable char- acter have been found so far only in the Dunvegan series.   The fuels

---

* See papers by the author, Trans. Royal Soc. of Canada, also reports of Dr. G. M. Dawson and Mr. Whiteaves.

† American Journ. of Science.

found in the area of the plains may be characterized generally as lignites, but on approaching the mountains these are found to contain a decreasing percentage of water, and eventually, in the foot-hills and areas included within the first limestone range, frequently become true coking bituminous coals, and in one instance, as above stated, have actually been converted into an anthracite which contains 86 per cent. of fixed carbon.

The Pleistocene deposits of this great interior region are of much interest, and are thus described by Dr. G. M. Dawson :—*

1. *Glacial Phenomena of the Laurentian Axis.*—Beginning with the glacial phenomena of the Laurentian axis, we may notice the appearances presented in the neighborhood of the Lake of the Woods only, where this axis is intersected by the forty-ninth parallel ; but, from the similarity of the traces of glacial action even in very distant parts of the Laurentian region, this will serve in some sense as a representation of its general features.

The Lake of the Woods, as a whole, occupies a depression in the south-western slope of the Laurentian region. It is over 70 miles in extreme length, and has a coast-line of between 300 and 400 miles. Its northern part is comparatively deep, reaching in some places a depth of over 80 feet. Its general form has been determined by that of an area of less highly altered rocks, which are probably Huronian ; and the details of its outline even follow very closely the changing character of the rock, spreading out over the schistose and thinly cleavable varieties, and becoming narrow and tortuous where compact dioritic rocks, greenstone conglomerate, and gneiss prevail. Its shores are almost invariably composed of solid rock with the rounded forms characteristic of ice-action, and dip rapidly below the surface of the water, forming a bold coast, sandy or gravelly beaches being comparatively rare. It is studded with innumerable islands, which vary from those several miles in length to mere water-wasted rocks. The islands, like the mainland, are seen, where not covered with luxuriant vegetation, to be composed of round-backed rocks. Only where the rocks are of a specially soft and schistose character has the action of the waters of the lake had sufficient effect on them to form cliffs. The southern part of the lake is very different : there are few islands ; the water is not deep ; and the whole southern shore is formed by low-lying deposits of sand and detrital matter. Where rock surfaces appear, however, they are like those of the northern part of the lake, heavily glaciated.

* Journal of Geol. Soc. of London, 1875.

All the harder rocks of the region still show with the utmost perfection the scratching and grooving of the glacial period; and some of the more compact granites and intrusive diorites retain a surface still perfectly bright and polished. On a small scale even the hardest and most homogeneous of the rocks show a tendency in the longer axis of their elevations to parallelism with the glacial markings. Though the general direction of the northern part of the lake also follows that of the ice-action, it is at the same time that of the belt of Huronian rocks already mentioned. The course of the glacial striæ is extremely uniform, and, from a great number of observations in different parts of the lake, is found to vary through a few points only, lying between north-north-east and south-south-west and north-east and south-west. Slight deflexions, sometimes observed, are generally traceable to deviation of the ice by masses of resistent rock running athwart its course, the striæ always showing a tendency to bend toward the more level regions, and away from the more elevated or rugged parts.

At a few places in the southern part of the lake, glaciation in the ordinary direction which gives form to the rock masses, was observed to be superinscribed with coarser scratches nearly east and west in direction. Some of these may be due to the packing of the ice of the lake itself in the spring; but instances occur which cannot be accounted for in this way. Some rock-surfaces on a low promontory in the southern part of the lake afford interesting examples. The most important direction, and that with which the forms of the surface coincide is here S. 13° W., superimposed on which at one place are scratches S. 45° W. or N. 45° E. Near this a direction of S. 50° W. or N. 50° E. occurs, on which is superimposed striation S. 15° W., a direction closely agreeing with the general one, and probably indicating a brief resumption of the original force after a short interval.

2. *Drift Plateau of Northern Minnesota and Eastern Manitoba.* The great plateau of Northern Minnesota, which stretches southward from the Lake of the Woods, shows only drift materials, and is composed of them to a great depth. Its general surface is remarkably uniform, and its slopes almost imperceptibly slight. It is, however, diversified on a small scale, being thickly strewn with shallow hollows, which are filled by little lakes or the almost impassable " muskegs " of the region. There are also low flat-topped ridges of sand and gravel of the nature of kames or eskers, and in many localities traces of larger lakes than those now existing, which have been drained by the gradual wearing down of the beds of their outfall streams.

The drift-deposits of this region rest on the gently sloping foot of the Laurentian axis, and are composed to a depth of 60 feet or more of fine sands and arenaceous clays, with occasional beds of gravel and small boulders. The finer deposits are generally very evidently false-bedded, and sometimes quite hard. The gravelly layers, as a rule, are found resting on the finer material between it and its surface soil, and sometimes lie on the denuded edges of the curved sand-beds below. On the Roseau River, about 30 feet from the top of the bank, a piece of wood protruded from a cliff of hard, sandy clay, and, on microscopic examination, appeared to be a fragment of the common cedar (*Thuja occidentalis.*) These distinctly-bedded deposits of the plateau appear to repose throughout on boulder-clay. This appears in the southern part of the Lake of the Woods; and on the Roseau River, also, indications of the underlying boulder-clay are found. In general, however, the few sections which exist do not penetrate sufficiently deep to show this deposit.

An interesting confirmation of the general direction already stated for the glacial action, is found in the composition of the materials of this plateau. Its eastern side, fronting on Lake Superior, is very abrupt, and seems to be held up by a ridge of hard old rocks, which here and there appears from beneath it. Ascending to the plateau-level from the extreme western point of Lake Superior, by the Northern Pacific Railway, the drift is seen to have a reddish-purple colour, which continues, though gradually becoming less marked, for some distance after attaining the summit. The colour then changes to the pale, yellowish gray which is generally characteristic of the drift of this plateau. The red drift is derived from the red rocks of the border of the lake, and is found along its whole southern side. It is here bounded by a line lying a short distance back from the north-western shore and nearly parallel to it. This western edge of the red drift has been already noticed by Whittlesey in his paper in the Smithsonian Contributions. The surface of the plateau is very generally strewn with erratics; and some of them are of great size. They are chiefly derived from the Laurentian and Huronian to the north; but there are also many of white limestone.[*]

3. *Lowest Prairie-Level and Valley of the Red River.* Descending the western side of the drift-plateau of Northern Minnesota, we enter the valley of the Red River. This trough runs nearly due north and south, and, from the south shore of Lake Winnipeg to the source of the Red River in Lake Traverse, is 315 miles in length. It does not end here,

[*] Bigsby, Journal Geological Society of London, 1851-52.

however, but passes by a continuous gap, never more than 690 feet above the sea-level, to the source of the Minnesota River, a tributary of the Mississippi. On the boundary-line the valley is 46 miles wide, and it narrows very gradually southwards. The floor of the valley, though it slopes upwards towards the sides, does so at so small an angle as to be quite imperceptible to the eye. It presents an appearance of perfect horizontality, and is, perhaps, the most absolutely level prairie-region of America. Looking down, towards evening, through one of the breaches in the edge of the western escarpment, it requires little imagination to suppose that the bluish level expanse is the sea; and, indeed, the whole of this valley must, at a time geologically modern, have been occupied by a great lake, the fine silty deposits of which now form its level floor. On examining these deposits they are found to be arranged in thin, horizontal beds, which together constitute a great thickness, and rest upon till or boulder-clay. Some of the layers immediately overlying the till may correspond with those already described in the same relative position on the drift-plateau; but nearly the whole thickness of the horizontal deposit probably belongs to the great lake of a later period. Stones of any kind are very seldom found on this prairie. They have no doubt been brought to their present position by the shore-ice of the lake itself, and are similar to those associated with the drift-deposits of its bounding escarpments.

Ascending the front of the western escarpment, it is found, as might almost have been foreseen, to be terraced; and on leaving the alluvial flat, boulders are again found abundantly, both strewing the terraces and the summit of the "mountain" or second prairie-steppe. The terraces not only occur on the front of this escarpment, but extend westward along the banks of the great valley of Pembina River, which at the time of their formation must have been an inlet of the lake, and is therefore probably of pre-glacial age.

4. *Second Prairie-Plateau.* The surface of the second plateau or steppe of the plain appears to be almost everywhere very thickly covered with drift deposits; and the undulations and slight irregularities of its contour, seem, in the main, due to the arrangement of these surface materials, which, though no doubt somewhat modified by subsequent denudation, do not seem to have suffered much. Over large areas no systems of "coulées" or stream-valleys are to be found; and the generally undulated surface must be due to original inequality of deposition, though a certain quantity of material has no doubt been removed from the rounded hillocks into the intervening basin-like swamps and hollows.

Such an arrangement not only implies the porous nature of the sub-soil, but is in accordance with the comparatively very small rainfall of the region, and would tend to show that at no time since its emergence has the precipitation been great. It was observed that in many places boulders and gravel are equally abundant on the crests of the gentle ridges and hillocks and in the hollows, while they are comparatively seldom seen on the intervening slopes. A similar observation has been made by Prof. Bell in a part of the second steppe considerably further north, and would tend to show slight erosion of the surface by marine currents subsequent to the deposition of the heavier materials.

The drift material is found generally to consist in great part of local debris derived from the immediately underlying soft formations; but this is always mixed with a considerable quantity of far-transported material, which is generally most abundant in the upper layers. Large erratics are in some localities very plentifully strewn over the plains, but they seem to be almost always superficial. They are generally of Laurentian rocks: but whitish and yellowish limestone, derived from the Silurian, flanking the western base of the Laurentian region, is abundant. A bank in Long-River Valley shows an interesting section, about 30 feet of drift, resting on Cretaceous clay or shale. Of the drift the lower portion is composed of stratified sands and gravels, which are evidently false-bedded. The pebbles are chiefly of the underlying rock, which, though soon splitting up under subaereal influences, has been hard enough to bear rounding under water. There are also a few examples of rocks of foreign origin, and the whole arranged in a manner implying a very strong flow of currents in different directions. About 11 feet from the top of the bank the false-bedded layers end abruptly, being cut off by a well-marked horizontal plane. Above this the bedding is nearly horizontal, and the drift includes many travelled boulders of Laurentian and white limestone, some of them large, together with much small Cretaceous stuff. Large boulders are also abundant, protruding from the surface of the prairie above.

In other places similar hard, yellowish, sandy clays are met with, but with little sign of stratification, holding many well glaciated stones, and thus resembling true till or boulder-clay.

With reference to the origin of the boulders and travelled stones the following table shows the proportions of different kinds of rock in a great number of samples taken from different parts of the second prairie steppe:

Laurentian	28.49
Huronian	9.71
Limestone	54.01
Quartzite Drift	1.14

The Laurentian material, consisting of granites and gneisses, are easily distinguishable. Those classed as Huronian are chiefly hard, greenish, epidotic, and hornblendic altered rocks. It is interesting to observe that the proportional importance of the Laurentian and Huronian, thus ascertained for the drift, is nearly that of their areas where they have been mapped. The limestone is that of the flanks of the Laurentian axis; and its great abundance is an interesting feature, and one tending to prove that this rock must in preglacial times have lapped far up on the Laurentian. These three classes are derived from the north-east or east. The fourth, or *Quartzite drift*, is a general name which may be applied to that coming from the Rocky Mountains, which, although not entirely composed of quartzite, is characterized by the great abundance of that material, and has a peculiar and distinctive appearance.

On the surface of this prairie-level there occur some remarkable elevated regions, which seem to be entirely composed of accumulated drift materials. The most prominent of these are included under the names of Turtle Mountain, Moose Mountain, and the Touchwood Hills. Though quite unconnected, these elevations follow in a general way a contour-line of the surface, and form a range roughly parallel to the Coteau, to which, in their appearance and material, they also bear the closest likeness. Of these elevations, the only one which I have personally examined is that known as Turtle Mountain, which is bisected by the forty-ninth parallel, and forms the most southern of the series. It is a region of broken hilly ground, which may be about 20 miles square, and is for the most part thickly wooded—a circumstance which renders it a specially prominent feature when viewed across the prairie. Its extreme height is not more that 500 feet above the prairie at its base ; and its general elevation is a little more than 2,000 feet above the sea, or nearly the same as that of the surface of the Coteau. On approaching it from the east, the already gentle-swelling plain becomes more markedly undulating, small basin-like swamps and ponds are more frequent, and its junction with the region of the " Mountain" would be undefinable but for the limiting border of the woods. The western end of the mountain is more abrupt towards the plain, and is much diversified with ridges, between which lie swamps and lakes, which show a general tendency to arrangement in north-and-south lines. Towards the eastern end there are somewhat extensive areas of gently undulating land, though always characterized by the abundance of pools and swamps. Notwithstanding the apparent abundance of water, there are few brooks or drainage-valleys, and the streams which do occur are quite small. The surface seems very nearly that of the drift as originally deposited, though sufficient fine material has been washed from the ridges to render the intervening hollows flat-bottomed.

5. *Edge of the Third Prairie-Plateau.*—One hundred and twenty miles west of Turtle Mountain the second prairie-plateau comes to an end against the foot of the great belt of drift deposits known as the Missouri Coteau. Beyond this point three diverse zones of country cross the forty-ninth parallel obliquely with a west-north-west course, in the order subjoined :—

1. Tumultuously hilly country based on a great thickness of drift, and forming the Coteau de Missouri properly so called.

2. Flat-topped *watershed plateau*, formed of rocks of the Lignite Tertiary, and constituting a part of the first tranverse watershed already described.

3. Lower, broken-down region, south of the plateau, partly based on the Lignite Tertiary, and characterized by gorges and large valleys draining towards the Missouri.

The second region can perhaps hardly be said to cross the line, but appears immediately north of it. On the line and southward the streams flowing to the Missouri rise near the southern edge of the first division, the greater part of the plateau having succumbed to denuding agencies.

The Missouri Coteau is one of the most important features of the western plains, and is certainly the most remarkable monument of the Glacial period now existing there. On the 49th parallel, the breadth of the Coteau, measured at right angles to its general course, is about 30 miles ; and it widens somewhat northward.

On approaching its base, which is always well defined at a distance, a gradual ascent is made, amounting in a distance of 25 miles to over 150 feet. The surface at the same time becomes more markedly undulating, as on nearing Turtle Mountain from the east, till, almost before one is aware of the change, the trail is winding among a confusion of abruptly rounded and tumultuous hills. They consist entirely of drift material ; and many of them seem to be formed almost altogether of boulders and gravel, the finer matter having been to a great extent washed down into the hollows and basin-like valleys without outlets with which this district abounds. The ridges and valleys have in general no very determined direction ; but a slight tendency to arrangement in north-and-south lines was observable in some places.

The boulders and gravel of the Coteau are chiefly of Laurentian origin, with, however, a good deal of the usual white limestone and a slight admixture of the quartzite drift. The whole of the Coteau-belt is characterized by the absence of drainage-valleys ; and in consequence its pools and lakes are often charged with salts, of which sulphates of soda and magnesia are the most abundant. The saline lakes frequently dry up completely towards the end of the summer, and present wide expanses of white efflorescent crystals, which contrast in colour with the crimson *Salicornia* with which they are often fringed.

Taking the difference of level between the last Tertiary rocks seen near the eastern base of the Coteau, and those first found on its western side, a distance of about 70 miles, we find a rise of 600 feet. The slope of the surface of the underlying rocks is therefore, assuming it to be uniform, a little less than 100 feet per mile. On and against this gently inclined plane the immense drift deposits of the Coteau hills are piled.

The average elevation of the Coteau above the sea, near the forty-ninth parallel, is about 2000 feet ; and few of the hills rise more than 100 feet above the general level.

Between the south-western side of the Coteau belt and the Tertiary plateau is a very interesting region with characters of its own. Wide and deep valleys with systems of tributary coulées have been cut in the soft rocks of the northern foot of the plateau, some of which have small streams still flowing in them fed by its drainage ; but for the most part they are dry, or occupied by chains of small saline lakes which dry up early in the summer. Some large and deep saline lakes also exist which do not disappear even late in the autumn. They have a winding, river-like form, and fill steep-sided valleys. These great old valleys have now no outlet ; they are evidently of preglacial age, and have formed a part of the former sculpture of the country. The heaping of the great mass of debris of the Coteau against the foot of the Tertiary plateau has blocked them up and prevented the waters finding their way northward as before ; and since glacial times the rainfall of the district has never been sufficiently great in proportion to the evaporation to enable the streams to cut through the barrier thus formed. The existence of these old valleys, and the arrangement of the drift-deposits with regard to them, throw important light on the former history of the plains.

Northward, the Coteau ceases to be identified with the Tertiary plateau, and rests on a slope of Cretaceous rocks. It can be followed by Palliser's and Hector's descriptions of the country to the elbow of the South Saskatchewan, and thence in a line nearly due north through the Eagle and Thickwood Hills ; beyond the North Saskatchewan, however, it appears to become more broken and less definite. In Dr. Hector's description of certain great valleys without outlet in this northern region, I believe I can recognize the existence of old blocked up river-courses similar to those just described. That the drift-deposits do not *form* the high ground of the watershed, but are merely piled upon it, is evident, as Cretaceous rocks are frequently seen in its neighborhood at no great depth.

South of the forty-ninth parallel the continuation of the belt of drift can be traced to a great distance southward and eastward in the territory of the United States, where it has been regarded as the terminal moraine of a hypothetical continental glacier.

In the Coteau, we have a natural feature of the first magnitude—a mass of glacial débris and travelled blocks with an average breadth of perhaps 30 to 40 miles, and extending diagonally across the central region of the continent for a distance of about 800 miles.

6. *Third or Highest Prairie-Plateau.*—Passing the Coteau and ascending the Plateau of the Tertiary, we notice at once a change in the character of the drift deposits. They are much thinner, and, area for area, perhaps do not equal one twentieth of those on the second prairie-steppe. They are also now largely composed of *quartzite drift* from the Rocky Mountains, of the nature of shingle, and seldom showing much trace of glaciation. With this western drift, however, a smaller proportion of that from the east or north-east is mingled. South of the watershed-plateau the third region (that sloping to the Missouri, where it is well sheltered to the north) shows the *quartzite drift* in even greater purity. Where, however, gaps or lower places in the watershed-plateau occur, incursions of Laurentian rocks and of eastern limestones are also found to a greater or less extent.

The general character of the travelled drift of the third steppe may be seen from its percentage composition, derived in the same way as already shown for the second steppe :—

Laurentian	27.05
Huronian	?
Limestone	15.84
Quartzite Drift	52.10

Though the percentage of Laurentian material appears nearly the same as before, the much smaller total quantity of drift on this level must be remembered. A mark of interrogation is put after Huronian, to indicate that a few specimens of this formation may be present, but, if so, are undistinguishable from some varieties of the *Quartzite drift*. The great decrease in limestone is at once seen ; and even the percentage here given includes some specimens of Rocky Mountain limestone which has travelled eastward with the *Quartzite drift*. The limestones of the flanks of the Laurentian were probably completely submerged ere the water reached the level of the third steppe. Quartzite and similar rocks now form over half of the entire travelled portion of the drift deposit.

Some of the lower parts of this steppe show thick deposits of true till or boulder clay, which holds in a hard yellowish sandy matrix well glaciated stones, both from the mountains and from the east, and also a great quantity of débris from the softer underlying beds, among which are fragments of lignite from the Tertiary. These deposits of till, though generally massive and weathering into rudely columnar forms in perpendicular banks, often show traces of bedding and arrangement in water ; and false-bedded sandy masses are found abruptly cut off above the confused bouldery clay. The shingle deposits of the higher levels may perhaps be formed partly from the rearrangement of this material ; they are at least superior to it.

The width of the third steppe, on the line, is about 450 miles ; but it narrows rapidly northward. Its surface is more diversified and worn than that of either of the other prairie-levels ; and the occurrence and features of the drift are less constant. Following it westward, and in the main slowly rising, Laurentian and Eastern limestone boulders continue to occur to within about 25 miles of the base of the Rocky Mountains, at a height of about 4200 feet. The distance of these travelled blocks from the nearest part of the Laurentian region is over 700 miles. Beyond this point eastern and and northern rocks were not found : but that the depression of the continent ceased here cannot be argued from this fact ; for by this time the whole of the Laurentian highlands would be submerged.

On the higher prairie, sloping up towards the mountains, the drift is entirely composed of material derived from them, and consists of quartzite, with softer shaly and slaty rocks, and limestone, which is generally distinguishable from that of the eastern origin. No granite or gneissic rocks occur in the vicinity of the forty-ninth parallel, or northwards in British America, in the eastern ranges, so far as is known. Southwards, in Montana, granites and gneisses are found underlying all the other formations, but they do not appear to be very extensively exposed.

7. *The Rocky Mountains.*—The brook issuing eastward from the mouth of the South Kootanie Pass has cut through a great thickness of clean gravel drift, composed of large and uniform well-rounded pebbles. Above

the brook, on the flanks of the mountains on the south side, are several well-preserved terrace-levels composed of similar material. The highest of these, though its altitude was not actually measured, was estimated from the known altitude of the Pass to be about 4400 feet above the sea. From the position of these terraces, in the open eastern throat of the pass, from which the whole surface of the country falls rapidly away, they can hardly be other than old seamarks. The topography of the region would not allow me to explain them on any hypothesis of a former moraine blocking up the valley.

Dr. Hector has measured similar terraces at several points along the Rocky Mountains north of the region now more especially under consideration, and states that they may be said to range from 3500 to 4500 feet above the sea. He also states that in the region examined by him the ordinary Laurentian erratics were not observed above 3000 feet, but mentions a very remarkable line of boulders of red granite deposited on the plains at a height of 3700 feet, which, knowing what we do now of the country, can hardly be supposed to have other origin than the Laurentian axis.

Among the Rocky Mountains themselves traces of the former action of glaciers are everywhere abundant, though in the part of the range near the forty-ninth parallel glaciers do not at present exist. They are, however, found further north. The evidence here met with so closely resembles that found in many other mountain regions as to render it unnecessary that it should be gone over in detail. Nearly all the valleys hold remnants of moraines, some of them still very perfect. The harder rocks show the usual rounded forms ; but striation was only observed in a single locality, and there coincided exactly with the main direction of the valley.

The valleys radiating from the summits of greatest elevation hold long lakes, many of which appear to be deep, and are filled with the most pellucid water. Whether they are in all cases dammed in by moraine matter I was unable to determine. These longer valleys very generally terminate in *cirques*, or amphitheatres, with almost perpendicular back and sides, which overlook small but deep terminal lakelets, held in by moraine-matter and shattered rock. In these sheltered hollows, and on the shady sides of the higher peaks, are masses of perennial snow, which have no doubt kept up the direct succession from the time when great *névés* filled the heads of the valleys and the mountains around them were completely snow-clad, and are only waiting some change in the climatic conditions to advance again down the old valleys and occupy the places they formerly filled.

8. *State of the Interior Region of the Continent previous to the Glacial Period.*—Before the onset of glacial conditions we find the continent standing at least at its present elevation, with its complete system of drainage from the larger river-valleys to many of their less important tributaries already outlined. Subaerial action must before this time have been in operation for a vast period, all the great features of the western plains having been already marked out, and the removal of a truly enormous mass of the soft and nearly horizontal Tertiary and Cretaceous rocks effected. That some very considerable changes in the direction of the drainage of the country in preglacial and in modern times took place, however, is probable. An examination of the Lake-of-the-Woods region and a comparison of levels render it almost certain that the waters of the area now drained by its tributary streams then found their outlet southward and westward, towards the present valley of the Red River, and that only after the blocking up of the southern region with the deposits of the drift did the waters flow over the pre-existing breach in the northern rim of the lake, and descend over the surface of the Laurentian to Lake Winnipeg. The Winnipeg River does not show any of the characters of a true river-valley, but consists of eroded and glaciated rock-hollows, from one to another of which the stream falls. There is also some evidence to show that the Red River itself, agreeing with the general structure of the country, flowed southwards; and if so, the Saskatchewan, too, would probably with it join the former representative of the Mississippi.

This subject, however, requires a more detailed discussion than can be granted it in this place.

9. *Mode of Glaciation and Formation of the Drift Deposits.*—To the precise manner in which the Glacial period was initiated, the area now in question gives no clue; but there does not seem to be either in the Laurentian region, or over the area of the plains, or in the Rocky Mountains, any evidence necessitating the supposition of a great northern ice-cap or its southward progress.

The great drift ridge of the Missouri Coteau at first sight resembles a gigantic glacier-moraine; and, marking its course in the map, it might be argued that the nearly parallel line of elevations, of which Turtle Mountain forms one, are remnants of a second line of moraine produced as a feeble effort by the retiring ice-sheet.

Such a glacier must either have been the southern extension of a polar ice-cap, or derived from the elevated Laurentian region to the east and north; but I think, in view of the physical features of the country, neither of these theories can be sustained.

To reach the country in the vicinity of the forty-ninth parallel a northern ice sheet would have to move up the long slope from the Arctic Ocean and cross the second traverse watershed, then, after descending to the level of the Saskatchewan valley, again to ascend the slope (amounting, as has been shown, to over 4 feet per mile) to the first transverse watershed and plateau of the Lignite Tertiary. Such an ice-sheet, moving throughout on broad plains of soft, unconsolidated Creta-ceous and Tertiary rocks, would be expected to mark the surface with broad flutings parallel to its direction, and to obliterate the transverse watersheds and valleys.

If it be supposed that a huge glacier resting on the Laurentian axis spread westward across the plains, the physical difficulties are even more serious. The ice moving southward, after having descended into the Red River trough, would have had to ascend the eastern escarpment of soft Cretaceous rocks forming its western side, which in one place rises over 900 feet above it. Having gained the second prairie-steppe, it would have had to pass westward up its sloping surface, and surmount the soft edge of the third steppe without much altering its form, and finally terminate over 700 miles from its source, and at a height exceed-ing the present elevation of the Laurentian axis by over 2,000 feet. The distribution of the drift equally negatives either of these theories, which would suppose the passage of an immense glacier across the plains.

In attributing the glacial phenomena of the great plain to the action of floating ice, we are in accord with Dr. Hector, who has studied a great part of the basin of the Saskatchewan—and also with Dr. Hayden, who, more than any other geologist, has had the opportunity of becoming familiar with all parts of the Western States.

The glaciating agent of the Laurentian plateau, in the Lake-of-the-Woods region, however, cannot have been other than glacier-ice. The rounding, striation, and polishing of the rocks there, are glacier-work; and icebergs floating, with however steady a current, cannot be supposed to have passed over the higher region of the watershed to the north, and then, following the direction of the striæ and gaining ever deeper water, to have borne down on the subjacent rocks. The slope of the axis, however, is too small to account for the spontaneous descent of ordinary glaciers. In a distance of about 30 miles, in the vicinity of the Lake of the Woods, the fall of the general surface of the country is only about 3½ feet to the mile. The height of the watershed region north-east of the lake has not been actually measured; but near Lac Seul, which closely corresponds with the direction required by glaciation, according to Mr. Selwyn's measurements it cannot be over 1400 feet. The height of land in other parts of the Laurentian region is very uniformly between about 1600 and 1200 feet. Allowing, then, 1600 feet as a maximum for the region north-east of the Lake of the Woods, and taking into account the height of that lake and the distance, the general slope is not greater than about 3 feet per mile—an estimate agreeing closely with the last, which is for a smaller area and obtained in a different way. This slope cannot be considered sufficient to impel a glacier over a rocky surface which Sir William Logan has well characterized as "mamillated," unless the glacier be a confluent one, pressed outwards mainly by its own weight and mass.

Such a glacier, I conceive, must have occupied the Laurentian highlands; and from its wall-like front were detached the icebergs which strewed the débris over the then submerged plains, and gave rise to the various monuments of its action now found there.

The sea, or a body of water in communication with it, which may have been during the first stages of the depression partly or almost entirely fresh, crept slowly upward and spread west-ward across the plains, carrying with it icebergs from the east and north. During its progress most of the features of the glacial deposits were impressed. In the section described at Long River we find evidence of shallow current-deposited banks of local material, afterwards, with deepening water, planed off by heavy ice depositing travelled boulders.

The sea reaching the edge of the slope constituting the front of the highest prairie-level, the deposition of the Coteau began, and must have kept pace with the increasing depth of the water and prevented the action of heavy ice on the front of the Tertiary plateau. The water may also have been too much encumbered with ice to allow the formation of heavy waves.

The isolated drift highlands of the second plateau, including the Touchwood Hills, Moose Mountain and Turtle Mountain, must also at this time have been formed. With regard to the two former, I do not know whether there is any preglacial nucleus round which drift-bearing icebergs may have gathered. There is no reason to suppose that Turtle Mountain had any such predisposing cause; but it would appear that a shoal once formed, by currents or otherwise, must have been perpetuated and built up in an increasing ratio by the grounding of the floating ice.

The Rocky Mountains were probably also at this time covered with descending glaciers; but these would appear to have been smaller than those of the Laurentian axis, as might, indeed, be pre-supposed from their position and comparatively small gathering-surface. The sea, when it reached their base, received from them smaller icebergs; and by these and the shore-ice the *quartzite drift* deposits appear to have been spread. That this material should have travelled in an opposite direction to the greater mass of the drift is not strange; for while the larger eastern and northern icebergs may have moved with the deeper currents, the smaller western ice may have taken directions caused by surface-currents from the south and west, or even been impelled by the prevailing winds. Some of the Laurentian débris, as we have seen, reached almost to the mountains, while some of the *quartzite drift* can be distinguished far out towards the Laurentian axis.

The occurrence of Laurentian fragments at a stage in the subsidence when, making every allowance for subsequent degradation, the Laurentian axis must have been far below water, would tend to show that the weight and mass of the ice-cap was such as to enable it to remain as a glacier till submergence was very deep. The Laurentian axis must always have been somewhat higher, and there may have been some inequality in the total amount of subsidence and elevation.

The emergence of the land would seem to have been more rapid. The water in retreat must have rearranged to some extent a part of the surface-materials. The quartzite drift of the third steppe was probably more uniformly spread at this time, and a part of the surface-sculpture of the drift-deposits of the second plateau may have been produced. It seems certain, however, that the Rocky Mountains still held comparatively small glaciers, and that the Laurentian region on its emergence was again clad to some extent with ice, for at least a short time. The closing episode of the Glacial period in this region, was the formation of the great fresh-water lake of the Red River valley, or first prairie-level (which was only gradually drained), and the re-excavation of the river courses.

It must not be concealed that there are difficulties yet unaccounted for by the theory of the glaciation and deposit of drift on the plains by icebergs; and chief among these is the absence, whereever I have examined the deposits and elsewhere over the west, of the remains of marine Mollusca or other forms of marine life. With a submergence as great as that necessitated by the facts it is impossible to explain the exclusion of the sea; for, besides the evidence of the higher western plains and Rocky Mountains, there are terraces between the Lake of the Woods and

Lake Superior nearly to the summit of the Laurentian axis, and corresponding beach-marks on the face of the northern part of the second prairie escarpment.

Conspectus of Geological Formations in the Manitoban Region, with some typical localities:—

QUATERNARY.	**Alluvium.** Lake deposits of Red River Valley and Peace River, &c.
	Stratified Sands and Gravels, and Moraines.
	Boulder Clay or Till. { Upper Boulder Clay. / Interglacial Lake Deposit. / Lower Boulder Clay. / Shingle Beds. } Of Southern Alberta, etc.
	Pliocene? Saskatchewan Gravels.
TERTIARY.	**Miocene.** Conglomerate Sandstone and Argillite of Cypress Hills, &c.

EOCENE, CRETACEO-TERTIARY, LARAMIE.	**Porcupine Hill Series.** / **Willow Creek Series.** / **St. Mary River Series.** (Of Southern Alberta.)	**Fort Union.** / **Laramie.** (Of Souris River, &c.)	**Wapite River Group.**
		Paskapoo Series. / **Edmonton Series.** } Northern Alberta	

CRETACEOUS.	**Fox Hill Series.** / **Pierre Series** / **Belly River Series.** / **Niobrara or Benton Series.** / **Kootanie Series.** (Of Alberta.)	**Fox Hill Series.** / **Pierre Series.** / **Niobrara Series.** / **Benton Series?** (Of Manitoba, &c.)	**Smoky River Group.** / **Dunvegan Group.** / **Fort St. John Group.** (Of Peace River.)

DEVONIAN.	Limestones of Manitoba Lake, etc.
SILURO-CAMB.	Trenton Group. { Limestones of Winnipeg Lake, Red River Valley, &c.
HURONIAN.	Between Red River Valley and Lake Superior.
LAURENTIAN.	West of Lake Winnipeg and Northward,

IV. THE BRITISH COLUMBIAN REGION.

The description of this region has, like the last, been taken mainly from the published memoirs of Dr. G. M. Dawson, F.G.S., with the exception of the Rocky Mountain region, the geology of which has been condensed from the Reports of R. S. McConnell, B.A.

From the western edge of the great plains to the Pacific, between the forty-ninth and fifty-sixth parallels, the Cordillera belt of the west coast has an average breadth of about 400 miles. Geologically, it may be considered throughout as a region of flexure and turmoil, and orographically, as one of mountains. As compared with its development in the Western States, the Cordillera belt may here be characterized as strict and narrow. To the north of the fifty-sixth parallel it has, as yet, only been traversed on a few lines and is very imperfectly known topographically and geologically. It appears probable, however, that a wide bay of comparatively undisturbed Cretaceous rocks may penetrate it in the region of the upper Liard River, which the echelon range bordering the lower part of this stream and the Mackenzie may bound to the east. A southern prolongation of such a range, beneath the newer rocks, may probably be indicated by the remarkable parallel flexures of the great rivers of the northern plains, the Liard, Peace, Athabasca and Saskatchewan.

Between the forty-ninth and fifty-sixth degrees of latitude the Cordillera belt is composed of four great ranges or axes of uplift and disturbance, which may be named the Rocky, the Gold, the Coast and the Vancouver Mountains. These are in the main nearly parallel and run in north-west and south-east bearings.

In its southern part, the Rocky Mountain range has an average breadth of about sixty miles, which decreases near the Peace River to forty miles or less. Near the forty-ninth parallel several summits occur with elevations exceeding 10,000 feet, but northward few attain this elevation till the head-waters of the Bow River are reached. About the sources of the North Saskatchewan and Athabasca the range appears to culminate, and Mounts Brown and Murchison occur with reputed altitudes of 16,000 and 13,500 feet respectively; near the Peace few summits exceed 6000, as far as known. Though more or less extensive snow-fields occur in many places, true glaciers appear only about the head waters of the Bow, North Saskatchewan and Athabasca. Where the line of junction has been closely examined, great faults occur along the line of boundary between the Mesozoic rocks of the eastern foot-hills and the palæozoic of the mountains, and numerous similar dislocations are found in the heart of the range parallel to its general course. With the exception of a single small area on the Upper Athabasca, reported by Dr. Hector, no crystalline schists have been found in this range, which consists almost entirely of sedimentary rocks, largely limestones. A few Cretaceous basins, resembling in their character isolated portions of the eastern foothills, are included in the southern part of the range. Some of the valleys penetrating the range on the east are lightly timbered or in part prairie-like in character, but, as a rule, these mountains are thickly wooded wherever sufficient soil exists for the support of trees, and, owing to the greater rainfall on the western slopes, the forests are there often very dense.

From the boundary line northward, the principal passes are as follows:—South Kootanie Pass, elevation, 7100 feet; North Kootanie Pass, eastern or main summit, 6750 feet; western summit, 6800 feet; Crow Nest Pass, summit, 5500 feet; Kanaskis Pass, 5700 feet; Kicking Horse Pass, 5300; Howse Pass, 5210 feet; Athabasca Pass, 6025 feet; Yellow Head Pass, 3733 feet; Smoky River Pass, 5300 feet; Pine River Pass, 2850 feet; Peace River valley, 2000 feet. With the exception of the route selected for the Canadian Pacific Railway, (the "Kicking Horse Pass,") these passes are traversed only by rough mountain trails, practicable for pack animals.

The western edge of the Rocky Mountain range is defined by a very remarkable straight and wide valley, which can be traced uninterruptedly from the forty-ninth parallel to the head waters of the Peace, a distance of 700 miles. This valley is occupied by the upper portions of several of the largest rivers, the Kootanie, Columbia, Fraser, Parsnip, and Findlay.

The Gold Ranges, which name may be applied as a general one to the next mountain region, are composed of a number of more or less clearly defined subsidiary ranges, the Selkirk, Purcell, Columbia and Cariboo Mountains. Crystalline schists, including gneisses and traversed by intrusive granitic masses, enter largely into the composition of these mountains, and there is ground for the belief that this is geologically the oldest of the ranges of this part of the Cordillera. Many of its summits exceed 8000 feet, and Mount Sir Donald in the Selkirks on the line of the Canadian Pacific Railway, is 10,645 feet high. Though many points on the culminating ridges of these mountains are abrupt and rugged in outline, extensive districts are occupied by rounded or plateau-like mountains, which contrast remarkably in appearance with the broken crags of the Rocky Mountains. The width of this mountain system may be stated as about eighty miles, but north of the Cariboo district, about the head waters of the Peace, it dies away completely, though probably again resuming in the Omenica district still further to the north-west.

Between this and the Coast Ranges, stretches a region which may be called the interior Plateau of British Columbia, with an average width of one hundred miles, and mean elevation of about 3500 feet. Its height, on the whole, increases to the south, while northward it falls gradually towards the group of large lakes, and the low country about the head waters of the Peace. It has over a great part of its area been covered by widespread flows of basalt and other volcanic rocks in the later Tertiary period, but is now dissected by deep and trough-like river valleys into most of which water standing at an elevation of 3000 feet above the present sea level would flow, dividing its surface into a number of islands. In some places the plateau is pretty level and uniform ; but usually it is only when broadly viewed that its character is apparent. It is practically closed to the north about latitude 55° 30' by the ends of several interlacated mountain ranges of which some of the summits attain 8000 feet. Nearly coinciding with the forty ninth parallel is a second transverse mountains zone, formed in a similar way, which may be considered as limiting it to the south, though traversed by several river valleys, of which that of the Okanagan is orographically the most important. The southern part of the plateau includes much open country, constitutes the best grazing region of British Columbia and offers besides some good agricultural land. To the north, with increasing moisture, it becomes generally forested.

The coast ranges, with an average width of one hundred miles, are frequently named the Cascade Mountains, but this term is a misleading one, as they are both geologically and orographically distinct from the well-known Cascade range of Washington Territory and Oregon. These mountains are largely composed of gneissic and granite rocks and crystalline schists. The average altitude of their higher peaks is between 6000 and 7000, while some exceed 9000 feet. Glaciers are of frequent occurrence and large size in their northern part, and on the Alaskan coast are known in several instances to descend to the sea level. These mountains are, as a rule, densely forested and extremely rugged, the flora of their seaward slopes being that characteristic of the west coast and co-ordinated with its excessive humidity, while on their northern and eastern flanks it resembles that of the inland ranges.

The name, Vancouver Range, may be applied to the fourth great mountain axis, which, in a partially submerged condition, constitutes Vancouver and Queen Charlotte islands, is continued to the south in the Olympian Mountains. The highest mountain of Vancouver Island reaches an elevation of 7484 feet, while there is a considerable mountainous area in the centre of the island which surpasses 2000 feet in average altitude. Several summits in the Queen Charlotte Islands exceed 4000 feet. This range, while still to a considerable extent formed of crystalline rocks like those of the Coast range, includes notable areas of stratified rocks, and is flanked in places by Cretaceous strata, important because of their coal-bearing character.

The most remarkable feature of the coast are its fjords and passages which, while quite analogous to those of Scotland, Norway, and Greenland, probably surpass those of any part of the world (unless it be the last-named country) in dimensions and complexity. The great height of the rugged mountain walls which border them also gives them a grandeur quite their own. The long river-like lakes of the interior of the province reproduce the features of these fjords on a smaller scale, and hold a homologous position to the inland ranges.

In treating of the rock structure of the Cordillera belt, in British Columbia, it will be most convenient to outline that of each of its great component regions, in so far as the older formations are concerned. The Cretaceous and Tertiary of the entire belt, which rest upon these in a comparatively little disturbed or altered state, may then be considered.

The ranges of the Rocky Mountains have been explored by Sir J. Richardson, Dr. Hector and Dr. Selwyn, and more recently by Dr. G. M. Dawson and Mr. R. G. McConnell.[*] According to the latter the whole belt may be divided into two regions, an eastern and a western. The former presents a series of faulted and often over-lapped rocks, indicating great pressure from the west and angular fractures and lateral thrusting of the beds. The latter is, on the other hand, a region of great folding without reversed faults and with much regional metamorphism.

The formations of the eastern belt are, in descending order :

1. The Cretaceous beds forming limited troughs.

2. The Banff limestones and shales and "Intermediate Limestone," which are Upper Palaeozoic (Devono-Carboniferous).

3. The Castle Mountain series, consisting of beds of Cambrian age.

The whole thickness of these beds is estimated at 18,000 feet.

In the western belt the highest beds are of Silurian age and from their characteristic coral have been named the *Halysites* beds. Below these are Siluro-Cambrian shales holding Graptolites *(Didymograptus, Glossograptus, Cryptograptus, Diplograptus, Climacograptus, etc.)* They are held by Lapworth to be of Trenton, Utica or Llandeilo age. Some of the forms are well-known American and European species. Under these are the Cambrian, represented by a great series of beds equivalent to those of the Castle Mountain series of the eastern belt and a lower series known as the Bow River group. The former consists largely of limestone and dolomite, the latter of argillite. It has afforded fossils of the genera *Paradoxides* and *Olenellus*, now regarded as of Middle or Lower Cambrian age.

---

[*] Geol. Survey of Canada, 1886.

The Cambrian fossils of the Rocky Mountains were first discovered by officers of the Geological Survey in 1884, and are noticed in McConnell's Report for 1887. Specimens have also found their way to the hands of United States geologists, and have been described by Romiger and Walcott, who, however, differ as to their classification.* According to Whiteaves the fossils belong to three distinct horizons. Lower and Middle Cambrian forms as *Olenellus* and *Paradoxides*, etc., were found near the base, and about 2000 feet higher such Middle and Upper Cambrian forms as *Olenoides* and *Doropyge* with two species of *Bathyurus*, while the upper beds yielded *Raphistoma rotuliformis* and an *Asaphus*, and are therefore probably Siluro-Cambrian.

The Cretaceous rocks of the eastern belt consist of shales holding in some places productive beds of Anthracite coal, and they are over-ridden by the older rocks with reversed faults and folded in abrupt troughs, showing that the great movements which have affected this belt are post-cretaceous.

South of the region just described, in the Rocky Mountains between the fiftieth and forty-ninth parallels, a wide-spread flow of contemporaneous, amygdaloidal trap occurs in the upper part of the section, and is probably referable to the Carboniferous period. Above this is a thickness of several hundred feet of red sandstones and flaggy magnesian sandstones and limestones.

These doubtless represent the Trias or Jura-Trias, which with similar lithological characters is very extensively developed in the Rocky Mountain region further south. The conditions indicated are those of an inland lake, and the occurrence of mud-cracks, ripple-marks, and the impressions of salt crystals show that considerable surfaces were at times dry, or but lightly covered by water. These beds have not been found further north in the range than the North Kootanie Pass, and it would appear probable that this is about the ancient limit of the Triassic inland sea. About the Peace River, and as far north as the Liard, Triassic beds have been found, but they are dark shales and sandstones quite different in character, and hold marine fossils of the age of the "Alpine Trias" of the Western States.

Unconformably underlying these are rocks which may now be provisionally classed as Cambrian, on the evidence of a few fossils found in the Columbia valley, and from analogy with beds described in the Western States since this section was first published, though it is quite

---

* Am. Journal of Science, 1888.

possible that Silurian beds may also be included. In some places, west of the Flathead River, the red beds included with these are characterized by sun-cracks, ripple-marks and prints of salt crystals precisely resembling those of the Trias and indicating similar conditions of deposit, though of vastly greater antiquity. Contemporaneous flows of diorite or diabase are also found at some horizons, and in tracing this series of rocks, which must in the aggregate be of great thickness, from point to point in the range, its lithological character is found to be very varied, and the subdivisions worked out in the neighbourhood of Watertown Lake, would appear to be inconstant.

In the Peace River district, on the fifty-fifth and fifty-sixth parallel, the axial mountains of the range are composed of massive limestones of Devonian and probably also of Carboniferous age, associated with saccharoidal quartzites. On the west side these are believed to overlie a series of argillites which occasionally become micaceous schists and slates, and also include quartzites. These rocks, which are probably Cambrian, are known to occupy a long trough east of the Parsnip river, and cross the Misinchinca with considerable width.

As before observed, the geological structure of the Gold Range, or second mountain axis, is little known. Cambrian rocks like those of the Rocky Mountains characterize its eastern portion, while its western side is largely composed of highly crystalline rocks, including gneiss, and considerable areas of granite. These are complicated, where they have been observed, by the occurrence of areas of much altered rocks resembling those of the interior plateau region, next to the west. Among these is a series of dark slates or schists which are the auriferous rocks in the Cariboo district and elsewhere. The age of these has not been determined.

The district coloured as Archæan on the published maps, while therefore probably in a large part composed of rocks of this period, is much more heterogeneous in character than can be indicated with our present information. The thickness of the crystalline rocks displayed on Shuswap Lake, has been estimated at about 32,300 feet. An isolated area of gneissic rocks doubtless belonging to a continuation of the main axis, as shown on the map, occurs at Carp Lake, west of McLeod. Lake.

The Palæozoic rocks of the interior plateau region of British Columbia are very varied in lithological character, but may be said, as a whole, to consist of massive limestone and compact or shaly quartzite, together with great beds of diorites or diabases, felspathic rocks and agglomerates, and some beds of serpentine. The last named material occurs in association with the contemporaneous volcanic materials, and doubtless represents the alteration product of olivine rocks. It is of interest as being of a period so recent as the Carboniferous. The limestones are not unfrequently converted to coarse-grained marbles, and together with the quartzite, appear in greatest force on the south-western side of the area they occupy. They have now been traced, maintaining their character pretty uniformly throughout, from the forty-ninth to the fifty-third parallel. Schistose or shaly argillite rocks, which may represent those folded with the Gold Range series, also occur.

In regard to the evidence of the age of these Palæozoic beds which must have in the aggregate a vast thickness, the following points may be stated. A portion at least of these rocks was, in 1871, shown by fossils collected by Dr. Selwyn, to belong to a horizon between the base of the Devonian and summit of the Permian; additional fossils have since been procured, of which the most characteristic is the peculiarly Carboniferous foraminifer *Fusulina*, which has now been found in several localities scattered over a wide area. While it is therefore quite possible that rocks of all stages of the Palæozoic may yet be identified in the interior plateau, there is as yet no proof that any of a date earlier than the Carboniferous occur, and the association of the other beds with the fossiliferous limestones is such as to show that a considerable part of them must be approximately of this age.

In the southern portion at least of the interior plateau region there exist, besides the Palæozoic rocks above described, and in addition to the probably in part Triassic argillites, extensive but as yet undefined areas of Triassic rocks of another character. These are largely of volcanic origin and have been designated the Nicola series. They have generally a characteristically green colour, but are occasionally purplish and consist chiefly of felspathic rocks and diabases, the latter often more or less decomposed. These rocks are in some cases quite evidently amygdaloidal or fragmental, and hold towards the base beds of grey sub-crystalline limestone intermingled in some places with volcanic material and containing occasional layers of water-rounded detritus.

Preliminary investigations along certain lines have now been made of the north-western extension of the region lying between the Rocky Mountains and the Coast Ranges in the extreme north of British Columbia and in the upper Yukon basin. Though its limiting ranges previously mentioned are still recognizable there, the orographic features of this country are not so simple as those in the south. The rocks, however, are closely analogous in general character to those above described, those of

Palæozoic age covering the greatest area. The Palæozoic rocks include limestones, quartzites, argillites, schists and slates with a notable proportion of agglomerates and other materials of volcanic origin, and are all pretty thoroughly altered and hardened and considerably flexed. Near Dease Lake, on the Pelly River, and elsewhere, beds of serpentine occur, and the associated rocks in these and many other places are preponderantly schistose and slaty, running through a number of varieties, but closely resembling the schistose and slaty rocks of Cariboo and other gold-bearing districts to the south, and here also yielding gold.

These Palæozoic rocks are here interrupted by granite areas, which generally rise in the form of ridges or mountain elevations, and are in some places flanked by more or less considerable occurrences of crystalline schists which appear to be more highly altered portions of the Palæozoic. Fossils of Carboniferous age have been obtained from the limestones in several places, and from shales in the Dease river graptolites of Middle Ordovician age have been determined.   Though the information available is still insufficient for the separation of the various formations, it is probable that rocks ranging from the Cambrian to the top of the Palæozoic and possibly also including the Triassic, may be embraced in this great preponderantly Palæozoic area.

The Coast Ranges of British Columbia are remarkably uniform in composition from the vicinity of the Fraser, where they originate, to the head of Lynn Channel—a length of about 900 miles. They are everywhere largely composed of grey granites which usually contain both hornblende and mica, and often show a notable abundance of the first-named mineral. With these are associated gneisses and other crystalline schists, together with belts of more or less highly altered argillites which often pass into black staurolitic and micaceous schists, while limestones are represented in some places by highly crystalline marbles. Some of these rocks are infolded and more or less completely altered portions of the Palæozoic strata previously described, but it yet remains to be determined whether there are represented any pre-palæozoic stratified rocks, or whether all those met with may be explained as products of the extreme metamorphism of palæozoic materials. Portions of the lower part of the Cretaceous series are found here and there at considerable elevations, resting in a comparatively unaltered state on the upheaved rocks of these ranges.

Vancouver and the Queen Charlotte Islands represent unsubmerged features of another mountain axis, but one of less energetic uplift. While granitic areas here still occur, these are of less importance and often manifestly intrusions.

In Vancouver and the Queen Charlotte Islands all the stratified rocks, of periods anterior to the Cretaceous, have suffered great flexure and disturbance, accompanied by more or less complete metamorphism, and this composition indicates throughout a continuance or recurrence of volcanic action on an enormous scale. Rocks originally of volcanic origin characterize the greater part of the entire area, and these in their present condition, at first sight, and as judged by eastern American analogies, might often be supposed to represent formations occupying a very low stage in the geological scale. These volcanic rocks, originally composed of minerals already crystalline, have yielded readily to subsequent alteration and recrystallization, and now, for the most part, appear as diabase, felsite and diorite, more or less massive, but passing locally into true schists. The massive varieties frequently show little evidence of a bedded character, but, when closely examined and traced out, are found to form portions of a stratified series of great thickness, which includes, besides the preponderant volcanic materials, important intercalations of flaggy and slaty argillites of quartzite and of limestone.

The greater part of the old volcanic rocks appear to have consisted originally of basaltic and trachytic lava flows, alternating with rough volcanic breccias and tuffs, largely composed of fragments of such flows. These have been deposited in the bed of a sea in which life was at intervals possible for considerable periods.

The entire series thus built up must have a thickness of many thousand feet, but no complete section of it can yet be given, and the evidences of age, as derived from fossils, is scanty. Obscure fossils, which are probably of Carboniferous age, have been found in some of the limestones of Vancouver Island, and beds of the same age may also occur in the southern part of the Queen Charlotte Islands. The most important zone of argillites and limestones, however, which is known in some places to attain a thickness of 2,500 feet, has yielded a number of fossils characteristic of the so-called Alpine Trias of California and the fortieth parallel region, which represents the Hallstadt and St. Cassian beds of Europe.

For this entire complex mass of pre-cretaceous rocks, the name Vancouver series is provisionally employed, though it is intended to restrict this name to the Triassic rocks when they shall have been distinctly separated.

Lying everywhere quite unconformably upon the older beds so far described, are the Cretaceous rocks, which constitute on the coast the true coal-bearing horizon of British Columbia. These rocks probably at

one time spread much more widely along the coast than they now do,
but have since been folded and disturbed during the continuation of the
process of mountain elevation, and have been much reduced by denudation.   Their most important area, including the coal-mining regions of
Nanaimo and Comox, may be described as forming a narrow trough
along the north-east border of Vancouver Island, 136 miles in length.
The rocks are sandstones, conglomerates and shales.   They hold abundance of fossil plants and marine shells in some places, and in appearance
and degree of induration much resemble the true Carboniferous rocks
of some parts of Eastern America.   In the Nanaimo area the formation
has been divided by Mr. J. Richardson as follows, in descending order :—

Sandstones, conglomerates and shales	3290 feet.
Shales......	660 "
Productive coal-measures.	1316 "
	5266 "

The last named consists of sandstones and shales, and holds valuable
coal-seams near its base.   In the Comox area seven well-marked subdivisions occur, constituting a total thickness of 4911 feet.

Upper conglomerate..........	330 feet.
Upper shales.	766 "
Middle conglomerate	1100 "
Middle shales............................	76 "
Lower conglomerate...	900 "
Lower shales........	1000 "
Productive coal-measures..........	739 "
	4911 feet.

The fuel obtained from these measures is a true bituminous coal, with
—according to the analysis of Dr. Harrington—an average of 6.29 per
cent. of ash and 1.47 per cent. of water.   It is admirably suited for most
ordinary purposes, and is largely exported, chiefly to San Francisco,
where, notwithstanding a heavy duty, it competes successfully with coals
from the west coast of the United States, owing to its superior quality.
The output of 1883 amounted to 213,000 tons, and is yearly increasing.

In addition to the main area of Cretaceous rocks above described, there
are numerous smaller patches, holding more or less coal, in different parts
of Vancouver Island, some of which may yet prove important.

In the Queen Charlotte Islands, Cretaceous rocks cover a considerable
area on the east coast, near Cumshewa and Skidegate Inlets.   At Skide-

gate they hold true anthracite coal, which, besides being a circumstance of considerable geological interest, would become, if a really workable bed could be proved, a matter of great economic importance to the Pacific coast.

At Skidegate, where these rocks are most typically developed, they admit of subdivision as follows, the order being, as before, descending :

A. Upper shales and sandstones .... 1500 feet.
B. Coarse conglomerates .... 2000 "
C. Lower shales with coal and clay ironstone. .. 5000 "
D. Agglomerates .... 3500 "
E. Lower sandstones.................... .... 1000 "

13,000 feet.

The total thickness is thus estimated at about 13,000 feet. With the exception of the agglomerates, the rocks in their general appearance and degree of induration compare closely with those of Vancouver Island. The agglomerates represent an important intercalation of volcanic material, which varies in texture, from beds holding angular masses a yard in diameter, to fine ash rocks, and appears at the junction to blend completely with the next overlying subdivision. The beds are generally felspathic and often more or less distinctly porphyritic.

At the eastern margin of the formation, the rocks lie at low angles, but become more disturbed as they approach the mountainous axis of the Islands, showing eventually in some cases overturned dips. It is in this disturbed region that the anthracite coal has been found, and from the condition of included woody fragments in the eastern portion of the area it is probable that any coal seams discovered there would be bituminous, like those of Vancouver Island.

The fossils from these beds are all from the lower portion of the formation, which is conclusively shown to represent the Chico group of the Californian geologists, which, with the locally developed Martinez group is considered to be equivalent to the Lower and Upper Chalk of Europe.* The highest subdivision of the Californian Cretaceous, the Tejon group, is supposed to represent the Maestricht, and, in the absence of fossils from the upper portion of the Vancouver Island formation, it is possible that it may be equally young. The flora of the Vancouver Cretaceous consists largely of modern angiospermous and gymnospermous genera, such as *Quercus, Platanus, Populus* and *Sequoia :* several of the genera and a few of the species being common to it and to the Dakota

---

* Whiteaves' Mesozoic Fossils, Reports Geol. Survey.

group of the Middle Cretaceous of the interior region of the continent. The botanical evidence, while yet imperfect, is therefore by no means in contradiction to that afforded by the animals and the stratigraphy.

A number of fossils from the Queen Charlotte Islands have also been described and figured from Mr. Richardson's collections and those of Dr. Dawson. There are few cases of specific identity between the forms in the Vancouver Cretaceous, previously described, and those of the Queen Charlotte Islands, the latter representing a lower stage in the Cretaceous formation. The plants found in these rocks, embracing numerous coniferous trees and a species of Cycad, also indicate a greater age than those of Vancouver.

The coal-bearing beds at Quatsino Sound, on the west coast of Vancouver Island, have also yielded a few fossils. These consist chiefly of well-characterized specimens of *Aucella Piochii*, which occurs but sparingly in the Queen Charlotte Islands, and brings the rocks into close relations with the Aucella beds of the mainland of British Columbia, and in Mr. Whiteaves' opinion probably indicate an " Upper Neocomian " age. The rocks of the Queen Charlotte Islands and Quatsino may therefore be taken together as representing the upper and lower portions of the so-called Shasta group of California, which in British Columbia can now be readily distinguished by their fossils.

On the mainland, developed most characteristically along the north-eastern border of the Coast Range, is a massive series of rocks first referred to by Dr. Selwyn, in the provisional classification adopted by him in 1871, as the Jackass Mountain group, from the name of the locality in which they are best displayed on the main waggon-road. The age of these rocks was not known at this time, but fossils have since been discovered in the locality above mentioned, and in several others, the most characteristic forms being *Aucella Piochii* and *Belemnites impressus*. The rocks are generally hard sandstones or quartzites, with occasional argillites, and very thick beds of coarse conglomerate. A measured section on the Skagit River includes over 4400 feet, without comprising the entire thickness of the formation. Behind Boston Bar, on the Fraser River, the formation is represented by nearly 5000 feet of rocks, while on Tatlyoco Lake it probably does not fall short of 7000 feet. At the last-named place these beds are found to rest on a series of felspathic rocks, evidently volcanic in origin, and often more or less distinctly porphyritic. On the Iltasyouco River, near the fifty-first parallel, and in similar relation to the Coast Range, an extensive formation characterized by rocks of volcanic origin, and often porphyritic, has also been found. Its thickness must be very great, and has been roughly estimated at one locality at 10,000 feet. It has been supposed, on lithological grounds, to represent the porphyritic formation of the vicinity of Tatlayoco Lake, and fossils found in it have been described as Jurassic. From analogy since developed with the Queen Charlotte Island fauna, however, Mr. Whiteaves now believes that the Iltasyouco beds are also Cretaceous.

Still further north the Cretaceous formation is not confined to the vicinity of the Coast Range, but spreads more widely eastward, being in all probability represented by the argillites and felspathic and calcareous sandstones of the Lower Nechacco ; and, as the explorations of 1879 have shown, occupying a great extent of country on the fifty-fifth parallel about the upper part of the Skeena and Babine Lake. They here include felspathic rocks of volcanic origin similar to those of the Itasyouco, which are most abundant on the eastern flanks of the Coast Range, and probably form the lower portion of the group. Besides these volcanic rocks, there is, however, a great thickness of comparatively soft sandstones and argillites, with beds of impure coal. The strata are arranged in a series of folds more or less abrupt, and have a general north-west or south-east strike. It is not impossible, from the general palæontological identity of the rocks of the interior with the older of those of the coast, that the Skeena region may eventually be found to contain valuable coal-seams, but this part of the country is at present very difficult of access, and there is no inducement to explore it.

More or less considerable areas of Cretaceous rocks are found still further northward, as far as the basin of the upper Yukon.

The Tertiary rocks do not form any wide or continuous belt on the coast of British Columbia, as is the case farther south. They are found near Sooke, at the southern extremity of Vancouver Island, in the form of sandstones, conglomerates, and shales, which are sometimes carbonaceous. Tertiary rocks also occupy a considerable area about the mouth of the Fraser River ; extending southward from Burrard Inlet, across the International boundary formed by the forty-ninth parallel, to Bellingham Bay and beyond. Thin seams of lignite occur at Burrard Inlet. Sections of the Tertiary rocks at Bellingham Bay are given in Dr. Hector's official report. Lignite beds were extensively worked here some years ago, but the mine has been abandoned owing to the superior quality of the fuels now obtained from Nanaimo and Seattle. About the estuary of the Fraser the Tertiary beds are much covered by drift and alluvial deposits, and are consequently not well known. Lignites, and even true coals, have been found in connection with them, but so far in beds too thin to

be of value. Fossil plants from Burrard Inlet and Bellingham Bay have
been described by Newberry and Lesquereux, and these are supposed to
indicate a Miocene age for the deposits.

Much farther north, in the Queen Charlotte Islands, the whole north-
eastern portion of Graham Island has now been shown to be underlaid
by Tertiary rocks, which produce a flat or gently undulating country,
markedly different from that found on most parts of the coast. The
prominent rocks are of volcanic origin, including basalts, dolerites,
trachytic rocks, and in one locality obsidian. Numerous examples of
fragmental volcanic rocks are also found. Below these, but seen in a
few places only, are ordinary sedimentary deposits, consisting of sandstones or shales, and hard clays with lignites. At a single locality on the
north end of Graham Island, beds with numerous marine fossils occur.
These, in so far as they admit of specific determination, represent shells
found in the later Tertiary deposits of California, and some of which are
still living on the north-west coast ; and the assemblage is not such as to
indicate any marked difference of climate from that now obtaining.

The Tertiary rocks of the coast are not anywhere much disturbed or
altered. The relative level of sea and land must have been nearly as at
present when they were formed, and it is probable that they originally
spread much more widely, the preservation of such an area as that of
Graham Island being due to the protective capping of volcanic rocks.
The beds belong evidently to the more recent Tertiary, and though the
palæontological evidence is scanty, it appears probable from this, and
by comparison with other parts of the west coast, that they should be
called Miocene.

To the east of the Coast or Cascade Range, Tertiary rocks are very
extensively developed. They have not, however, yielded any marine
fossils, and appear to have been formed in an extensive series of lakes,
which may at one time have submerged much of the region described as
the interior plateau. The Tertiary lakes may not improbably have been
produced by the interruption of the drainage of the region by a renewed
elevation of the coast mountains, proceeding in advance of the power of
the rivers of the period to lower their beds ; the movement culminating
in a profound disturbance leading to a very extensive volcanic action.
The lower beds are sandstones, clays, and shales, generally pale-greyish or
yellowish in colour, except where darkened by carbonaceous matter.
They frequently hold lignite, coal, and in some even true bituminous
coal occurs. The sedimentary beds rest generally on a very irregular
surface, and consequently vary much in thickness and character in differ-

ent parts of the extensive region over which they occur. The lignites appear in some places to rest on true "underclays," representing the soil on which the vegetation producing them has grown, while in others—as at Quesnel—they seem to be composed of drift-wood, and show much clay and sand interlaminated with the coaly matter.

In the northern portion of the interior the upper volcanic part of the Tertiary covers great areas, and is usually in beds nearly horizontal, or at least not extensively or sharply folded. Basalts, dolerites, and allied rocks of modern aspect occur in sheets, broken only here and there by valleys of denudation, and acidic rocks are seldom met with except in the immediate vicinity of the ancient volcanic vents. On the lower Nechacco, and on the Parsnip River, the lower sedimentary rocks appear to be somewhat extensively developed without the overlying volcanic materials.

The southern part of the interior plateau is more irregular and mountainous. The Tertiary rocks here cover less extensive areas, and are much more disturbed, and sometimes over wide districts—as on the Nicola—are found dipping at an average angle of about thirty degrees. The volcanic materials are occasionally of great thickness, and the little disturbed basalts of the north are, for the most part, replaced by agglomerates and tufas, with trachytes, porphyrites, and other felspathic rocks. It may indeed be questioned whether the character of these rocks does not indicate that they are of earlier date than those to the north, but, as no direct palæontological evidence of this has been obtained, it is presumed that their different composition and appearance is due to unlike conditions of deposition and greater subsequent disturbance. No volcanic rocks or lava flows of Post-glacial age have been met with.

The organic remains so far obtained from these Tertiary rocks of the interior consist of plants, insects, and a few fresh-water molluscs and fish scales, the last being the only indication of the vertebrate fauna of this period. The plants have been collected at a number of localities. They have been subjected to a preliminary examination by Sir W. Dawson, and several lists of species published. While they are certainly Tertiary, and represent a temperate flora like that elsewhere attributed to the Miocene, they do not afford a very definite criterion of age, being derived from places which must have differed much in their physical surroundings at the time of the deposition of the beds. Insect remains have been obtained in four localities. They have been examined by Mr. S. H. Scudder, who has contributed three papers on them to the Geological Reports, in which he describes forty species, all of which are considered

new.   None of the insects have been found to occur in more than a
single locality, which causes Mr. Scudder to observe that the deposits
from which they came may either differ considerably in age, or, with
the fact that duplicates have seldom been found even in the same locality,
evidence the existence of different surroundings, and an exceedingly rich
insect fauna.

Though a large part of the interior plateau may at one time have been
pretty uniformly covered with Tertiary rocks, it is evident that some
regions have never been overspread by them, while, owing to denudation,
they have since been almost altogether removed from other districts, and
the modern river valleys often cut completely through them to the older
rocks.   The outlines of the Tertiary areas are therefore now irregular
and complicated.

Pleistocene Deposits.—The following notes embrace the latest observations of Dr. G. M. Dawson, in British Columbia, which seems to be
remarkable for the great development of its glacial phenomena.—

It has been shown* that at one stage in the Glacial period—that of maximum
glaciation—a great confluent ice-mass has occupied the region which may be named the
Interior Plateau, between the Coast Mountains and Gold and Rocky Mountain
Ranges. From the fifty-fifth to the forty-ninth parallel this great glacier has left
traces of its general southward or south-eastward movement, which are distinct from
those of subsequent local glaciers. The southern extensions or terminations of this
confluent glacier, in Washington and Idaho Territories, have quite recently been
examined by Mr. Bailey Willis and Prof. T. C. Chamberlin, of the U.S. Geological
Survey.† There is, further, evidence to show that this inland ice flowed also, by transverse valleys and gaps, across the Coast Range, and that the fiords of the coast were
thus deeply filled with glacier-ice, which, supplemented by that originating on the
Coast Range itself, buried the entire great valley which separates Vancouver Island
from the mainland, and discharged seaward round both ends of the island. Further
north, the glacier extending from the mainland coast, touched the northern shores of
the Queen Charlotte Islands. The observed facts on which these general statements
are based have been fully detailed in the publications already referred to, and it is not
the object of this note to review former work in the region further than to enumerate
the main features developed by it, and to add to these a summary of observations made
during the summer of 1887 in the extreme north of British Columbia, and in the Yukon
basin beyond the sixtieth parallel, which forms the northern boundary of that province.

The littoral of the south-eastern part or "coast strip" of Alaska presents features
identical with those of the previously examined coast of British Columbia, at least as
far north as lat. 59°, beyond which I have not seen it. The coast archipelago has
evidently been involved in the border of a confluent glacier which spreads from the
mainland and was subject to minor variations in direction of flow dependent on surface
irregularities, in the manner described in my report on the northern part of Vancouver

---

* Quart. Journ. Geol. Soc. vol. xxxi. p. 89. *Ibid.* vol. xxxiv. p. 272. Canadian
Naturalist, vol. viii.

† Bulletin U.S. Geol. Survey, No. 40, 1887.

Island.* No conclusive evidence was here found, however, either in the valley of the Stikine River or in the pass leading inland from the head of Lynn Canal, to show that the ice forced seaward across the Coast Range, though analogy with the coast to the south favours the belief that it may have done so. The front of the glacier must have passed the outer border of the archipelago, as at Sitka, well-marked glaciation is found pointing toward the open Pacific † (average direction about S. 81° W. astr.) It is, however, in the interior region, between the Coast Range and the Rocky Mountains proper and extending northward to lat. 63°, explored and examined by us in 1887, that the most interesting facts have come to light respecting the direction of movement of the Cordilleran glacier. Here, in the valleys of the Pelly and Lewes branches of the Yukon, traces were found of the movement of heavy glacier-ice in a northerly direction. Rock-surfaces thus glaciated were observed down the Pelly to the point at which it crosses the 136th meridian, and on the Lewes as far north as lat. 61° 40', the main direction in the first-named valley being north-west, in the second north-north-west. The points referred to are not, however, spoken of as limiting ones, for rock exposures suitable for the preservation of glaciation are rather infrequent on the lower portions of both rivers, and more extended examination may result in carrying evidence of the same kind much further toward the elevated plains of the lower Yukon. Neither the Pelly valley nor that of the Lewes is hemmed in by high mountainous country except towards the sources, and while local variations of the kind previously referred to are met with, the glaciation is not susceptible of explanation by merely local agents, but rather implies the passage of a confluent or more or less connected glacier over the region.

In the Lewes valley, both the sides and summits of rocky hills 300 feet above the water were found to be heavily glaciated, the direction on the summit being that of the main (north-north-west) orographic valleys, while that at lower levels in the same vicinity followed more nearly the immediate valley of the river, which here turns locally to the east of north.

Glaciation was also noted in several places in the more mountainous country to the south of the Yukon basin, in the Dease and Liard valleys, but the direction of movement of the ice could not be determined satisfactorily, and the influence of local action is there less certainly eliminated.

Of the glacial deposits with which the greater part of the area of the inland region is mantled it is not intended here to give any details, though it may be mentioned that true Boulder-clay is frequently seen in the river sections, and that this generally passes upward into, and is covered by important silty beds, analogous to the silts of the Nechacco basin, further south in British Columbia, and to those of the Peace River country to the east of the Rocky Mountains. It may be stated also that the country is generally terraced to a height of 4000 feet or more, while on an isolated mountain-top near the height of land between the Liard and Pelly rivers (Pacific-Arctic watershed) rolled gravel of varied origin was found at a height of 4300 feet, a height exceeding that of the actual watershed by over 1000 feet.

Reverting to the statements made as to the direction of the general glaciation, the examination of this northern region may now be considered to have established that the main gathering-ground or névé of the great Cordilleran glacier of the west coast, was included between the fifty-fifth and fifty-ninth parallels of latitude in a region which, so far as explored, has proved to be of an exceptionally mountainous character.

---

* Annual Report Geol. Survey, Canada, 1885, p. 100 B.

† Mr. F. G. Wright has already given similar general statements with regard to this part of the coast of Alaska, (American Naturalist, March, 1887.)

It would further appear that this great glacier extended, between the Coast Range and the Rocky Mountains, south-eastward nearly to lat. 48°, and north-westward to lat. 63°, or beyond, while sending also smaller streams to the Pacific Coast.

In connection with the northerly direction of ice-flow here mentioned, it is interesting to recall the observations which I have collected in a recently published report of the Geological Survey, relating to the northern portion of the continent east of the Mackenzie River.* It is there stated that for the Arctic coast of the continent, and the islands of the archipelago off it, there is a considerable volume of evidence to show that the main direction of movements of erratics was *northward*. The most striking facts are those derived from Prof. S. Houghton's Appendix to M'Clintock's Voyage, where the occurrence is described of boulders and pebbles from North Somerset, at localities 100 and 135 miles north-eastward and north-westward from their supposed points of origin. Prof. Haughton also states that the east side of King William's Land is strewn with boulders of gneiss like that of Montreal Island, to the southward, and points out the general northward ice-movement thus indicated, referring the carriage of the boulders to floating ice of the Glacial Period.

The copper said to be picked up in large masses by the Eskimo, near Princess Royal Island, in Prince of Wales Strait, as well as on Prince of Wales Island,† has likewise, in all probability been derived from the copper-bearing rocks of the Coppermine River region to the south, as this metal can scarcely be supposed to occur in place in the region of horizontal limestone where it is found.

Dr. A. Armstrong, Surgeon and Naturalist to the "Investigator," notes the occurrence of granitic and other crystalline rocks not only on the south shore of Baring Land, but also on the hills at some distance from the shore. These, from what is now known of the region, must be supposed to have come from the continental land to the southward.

Dr. Bessels, again, remarks on the abundance of boulders on the shore of Smith's Sound in lat. 81° 30′, which are manifestly derived from known localities on the Greenland coast much farther southward, and adds, " Drawing a conclusion from such observations, it becomes evident that the main line of the drift, indicating the direction of its motion, runs from south to north.‡

It may further be mentioned that Dr. R. Bell, of the Canadian Geological Survey, has found evidence of a northward or north-eastward movement of the glacier-ice in the northern part of Hudson Bay, with distinct indications of eastward glaciation in Hudson Strait.§ For the northern part of the great Mackenzie Valley we are as yet without any very definite information, but Sir J. Richardson notes that Laurentian boulders are scattered westward over the nearly horizontal limestones of the district.

Taken in conjunction with the facts for the more southern portion of the continent, already pretty well known, the observations here outlined would appear to indicate a general movement of ice outward, in all directions, from the great Laurentian axis or plateau which extends from Labrador round the southern extremity of Hudson Bay to the Arctic Sea; while a second, smaller, though still very important region of dispersion—the Cordilleran glacier-mass—occupied the Rocky Mountain region on the west, with the northern and southern limits before approximately stated.

---

* Notes to accompany a Map of the Northern Portion of the Dominion of Canada, east of the Rocky Mountains, p. 57 R., Annual Report, 1886.

† De Rance, in Nature, vol. xi, p. 492.

‡ Nature, vol. ix.

§ Annual Report Geol. Surv. Canada, 1885, p. 14 D.D. ; and Report of Progress, 1882-84, p. 36 D.D.

The gold of British Columbia has long been known and worked in placer deposits along the gold or second mountain range, at Cariboo, Cassiar, and elsewhere.   The original site of this gold is in the slaty rocks of this range.   Silver ores and native silver are found at various places in British Columbia (Fort Hope, Cheny Creek, etc.) The Cretaceous coal of Naniamo, in Vancouver Island, is extensively mined and beds of Anthracite have been opened in the Cretaceous troughs of the Rocky Mountains.

	COAST REGION.		INTERIOR REGION.	ROCKY MOUNT. REGION
Quaternary.	Recent Raised Beaches. Stratified Sands, Gravels and Clays (Marine shells). Boulder Clay or Till.		Stratified Sands and Gravels, "Wh. Silts" of Nechacca, etc. Terrace Deposits, Moraines and Till.	
Tertiary.	Pliocene (?) Pre-glacial gravels (often Auriferous). Miocene (Volcanic). Miocene (Sedimentary, Marine Shells.)		Miocene (Volcanic). Miocene (Sedimentary, with lignites).	Shales and Sandstones of the Flathead rivers.
Cretaceous	*Nanaimo Basin.* Sandst.　329' 1326' Productive Coal Meas. 739' Chico of Cal. *N. Part of Vancouver Island.* (Quotsins, etc.) A. Up. Shales B. Conglomerates. C. L. Sandst. & Shales (with coal) D. Wanting. E. Wanting. Shasta of Cal.	*Comox Basin* Up. Cong.　320' Up. Shales 776' Mid. Cong. 1100' Mid. Shales 76' L. Cong.　900' L. Shales 1000' *Queen Charlotte Islands.* A. U. Shales & Sandst. 1500' B. Conglomerates.　2000' C. L. Shales & Sandst. 5000' (with coal) D. Agglomerate.　3500' E. L. Sandstones. 1000'		St. Mary River Series (base) Fox Hill and Pierre. Belly River Series. Benton & Niabrara. 1400 Volcanic rocks (local) 2750 Dakota ⎱ Kootanie ⎰ 7000' Iltasyouco Beds 10,000' Aucella Beds of Tatlayoco, Jackass Mt. and Skagit 7000 also rocks on the Lewes, Skeena and Nechacco rivers and elsewhere, often including volcanic materials.
Triassic.	Vancouver Series (largely volcanic, but with Monotis shales and limestones.		Nicola Series (largely Volcanic, but with Monotis shales and limestones	Monotis Beds of Northern Rocky Mts. Red Beds of S. Rocky Mts.
Palæozoic.	*(Carboniferous.)* Altered Volcanic Rocks and Limestones (these have not yet been distinctly separated from the Vancouver Series).		*(Carboniferous).* Cache Creek Group. (Limestones, quartzites and altered volcanic rocks). Altered Volcanic rocks, limestones, slates and quartzites (becoming schistose locally. Probably referable to various horizons of the Palæozoic, but not yet sub-divided). *(Cambrian).* Quartzites, Slates, etc., of Purcell & Selkirk ranges.	*(Carboniferous).* Upper Banff Shales. Up. Banff Limestones. Lower Banff Shales. Low. Banff Limestones. *(Devonian).* Intermediate Limest. *(Silurian).* Halysites Beds. *(Cambro-Silurian.)* Graptolitic Shales. Castle Mt. Limestone, (Upper part). *(Cambrian).* Castle Mt. Limestone. (Lower part). Bow River Group.
Arch'n.	Portions of Basal Rocks of Coast Ranges (?).		Gneissic & Schistose rocks of Shuswap L. & Gold Range.	

V. THE ARCTIC OR HUDSONIAN REGION.

This has as yet been only partially surveyed by the officers of the Canadian Survey, principally by Dr. R. Bell, and our information respecting it is largely due to the observations made and specimens and facts collected by the various exploring expeditions. The following summary is chiefly derived from Dr. Dawson's "Notes to accompany a map of the Northern portion of Canada," 1887.

The region includes what may be termed the Canadian portion of the Basin of the Arctic Sea, extending from Greenland, Baffin Land and Labrador on the east to Alaska on the west. It consists in great part of the northern portion of the great Archæan nucleus of the continent with belts of Palæozoic and later rocks around Hudson Bay and the Arctic Sea, and in the valley of the Mackenzie River.

1. *General Geological Structure.*—The Archæan or Eozoic rocks are dominant in the northern part of the continent. They form also, so far as has been ascertained, the greater part of Greenland, and doubtless underlie, at no great depth, the entire Arctic archipelago. While the information available is sufficient to indicate the existence of the different subdivisions of the Archæan which are met with in the southern portion of Canada, including the lowest Laurentian or granitoid gneiss series, the Middle Laurentian, possibly the peculiar rocks classed as the "Upper Laurentian," and certain of the more schistose and generally darker coloured and more basic rocks classed as Huronian, it is far too imperfect to admit of the separation of these subdivisions on the map. It is evident that the Huronian is represented in parts of the west coast of Greenland, and it is probably also recognizable on the Labrador coast, and on the west coast of Hudson Bay, and possible that it is elsewhere present over the Archæan area, in proportions as great as it has been found to hold where these rocks have been subjected to more systematic and detailed investigation. The distribution of the Huronian is important from an economic point of view, on account of its generally metalliferous character, which may eventually give value to tracts of country in which the rigorous nature of the climate entirely precludes the possibility of agriculture.

While the term Cambrian may be interpreted in the widest sense, namely, as including all rocks above the Huronian, to the base of the Cambro Silurian of the Canadian Geological Survey, and the reference of the rocks of the region here treated of to the Cambrian is based entirely on lithological and stratigraphical grounds, the rocks so classed are, as far as known, probably referable to the Lower Cambrian. It is further quite

evident that in the extensive area which has been coloured on the map as
Cambrian on the Arctic coast, in the vicinity of the Coppermine River,
the rocks are analogous in character to those of the Kewenaw or Animikie
of the Lake Superior region, and probably represent both groups of that
great copper-bearing series. The mere occurrence of native copper in
considerable quantities on the Coppermine, in association with prehnite
and other minerals resembling those which accompany it on Lake Superior,
gives a *prima facie* probability to this correlation, which is borne out by
a more careful study of Sir J. Richardson's accurate notes, and was
recognised by Richardson himself, who had examined both regions. Prof.
R. D. Irving states that the Animikie of Hunt, or "Lower Group" of
Logan, on Lake Superior, is composed of a great thickness of quartzites,
quartz-slates, argillaceous or clay-slates, magnetic quartzites and sandstones, thin limestone beds, and beds of a cherty or jaspery material,
associated with coarse gabbro and fine grained diabase (Copper-Bearing
Rocks of Lake Superior, 1883, p. 379,) while the overlying Kewenaw
series is made up of similar basic crystalline rocks, with interbedded
detrital rocks, chiefly reddish conglomerates and sandstones, the conglomerates consisting, for the most part, of pebbles of acidic crystalline
rocks. A comparison of the above description with that of Richardson of
the rocks of the Coppermine, shows the practical lithological identity of
the two widely separated areas.

Though not a geologist, Captain Back, who had seen the Coppermine
rocks, referred the formation coloured as Cambrian on Great Slave Lake,
to the same series, from its lithological similarity ; to which also the
doubtfully placed area of Cambrian on Back's route from Great Slave
Lake to the mouth of Great Fish River is attributed. To this formation also may be referred the great volcanic series, described by Dr. R.
Bell as the Manitounuck group, of the east coast of Hudson Bay, and
the red sandstones of his "Intermediate group," which he regards as
unconformably underlying the Manitounuck rocks, may possibly also belong to the Keewenaw or Animikie.

Throughout the whole of the vast northern part of the continent, this
characteristic Cambrian formation, composed largely of volcanic rocks,
apparently occupies the same unconformable position with regard to the
underlying Laurentian and Huronian systems. Its present remnants
serve to indicate the position of some of the earliest geological basins,
which, from the attitude of the rocks, appear to have undergone comparatively little subsequent disturbance. Its extent entitles it to be recognized as one of the most important geological features of North America.

The Silurian and Cambro-Silurian (Upper and Lower Silurian) rocks are chiefly pale limestones, often of a yellowish or cream colour and frequently more or less dolomitic. They rest everywhere unconformably on the Archæan or on the Cambrian rocks, and one of their most constant features appears to be the existence of a zone of red sandstones or arenaceous limestones and conglomerates at the base, a fact which leads us to suspect that the red sandstones of Tuunudleorbick and Igalliko in Greenland, which have been doubtfully referred on lithological grounds to the Permian, Devonian and Cambrian, may belong in reality to the Silurian.

This great Silurian and Cambro-Silurian limestone series is very widely developed, and is, in most places, nearly horizontal and undisturbed, with long light undulations in the bedding or persistent and uniform dips at very low angles. These features are very prominently shown in the sketches of many parts of the coast line in the Arctic islands, reproduced in the volumes of voyagers. But for the undisturbed and flat condition of the limestones, and the formations overlying them in the Arctic basin, it would be impossible, with the fragmentary geological information available, to offer any proximately correct geological map of the region as a whole.

In a paper printed in the report of the British Association for 1855, J. W. Salter states that the Silurian fossils, obtained up to that time, showed a uniform horizon of Upper Silurian limestone, stretching from near the entrance of Barrow Strait to Melville Island and far to the south along Prince Regent Inlet, and argues therefrom a wide extent of circumpolar land in Lower Silurian (Cambro-Silurian) times. In this he was followed, two years later, by Sir R. Murchison, who writes:—"I am led to believe that the oldest fossiliferous rock of the Arctic regions is the Upper Silurian." (Appendix to McClure's voyage, p. 402 ; Siluria, p. 440.) Though the Upper Silurian beds undoubtedly occupy a great part of the American polar region, characterizing the "south of North Devon and nearly all the islands south of Melville and Lancaster sounds, including the south of Banks Land, Prince of Wales Land, King William Land, North Somerset, Boothia Felix, etc." (Fielden and De Rance, Quart. Journ. Geol. Soc. vol. xxxiv.), the occurrence of Lower Silurian (Utica) fossils in Frobisher Bay, as shown by Hall's collections, on the shores of Kennedy Channel, as determined by Etheridge, and the occasional discovery of Lower Silurian forms in the regions above, referred in a general way to the Upper Silurian, prove that the generalization made by Salter and Murchison, on the evidence of less complete collections, can-

not now be admitted, and that the limestones of the Arctic represent probably the whole of the Siluro-Cambrian and Silurian, and possibly part
of the Devonian.   (See Fielden and De Rance, *loc. cit.*)   Heer enumer-
ates the following places, besides those above particularly referred to, as
yielding Lower Silurian types :—North Devon, Cornwallis Island, Griffith
Island, west coast of King William Land, Boothia (Flora Fossilis Arctica
vol. i., p. 24.)

The above allusion to the possibly Devonian age of part of the limestones of the Arctic basin proper, is of interest in connection with the
question of the relation of these limestones to the equally important limestone series of the Mackenzie River Region.   The early reference of an
extensive portion of these latter to the Silurian by Isbister and others,
can scarcely now be maintained, since Meek, as the result of his examination of the most ample collection of fossils which has ever been brought
together from the Mackenzie valley, reports the existence in the limestones of Devonian forms alone, though, as he cautiously remarks, he is
not prepared to deny the existence of Silurian rocks.   This Devonian
facies is maintained by the limestones of the Mackenzie valley to the very
shores of the Arctic Sea, as shown by the occurrence of Hamilton group
fossils on the Anderson River.

Rocks of the Lower Carboniferous or so-called "Ursa Stage" are
widely distributed in the Arctic Archipelago, and their character, as described by Prof. Haughton from an examination of the specimens brought
back by voyagers is similar to that of these rocks in Spitzbergen, as described by Heer, and to the Horton series of Nova Scotia.   This formation,
which both from its extent, and in its character as a coal-bearing series is
a very important one, should apparently be regarded throughout as Lower
Carboniferous and equivalent to the Tweedian of the north of England
and of Scotland, and to the Culm of Germany.

It may be noted that we are as yet without the data for any accurate
estimate of the entire thickness of these or the previously mentioned
rock-series of the Arctic basin.

Certain small outlying areas in the northern part of the Arctic Archipelago have been referred to the Lias.   These it appeared possible might
now be assigned to the "Alpine Trias," a formation which since the above
reference was made has been found to be wide-spread and important in
the Cordillera region of North America as far north as the northern part
of British Columbia, and is also characteristically developed in Spitzbergen
and the north-west of Siberia.   This question was referred to Prof. S.
Haughton, who had originally described the fossils on which the age of

the beds in question had been determined. The result of a critical re-examination of the fossils, which Prof. Haughton was so kind as to have made, appears, however, rather to confirm the original Liassic or Jurassic reference of these northern rocks.

The Tertiary rocks of the Mackenzie River and of Greenland, so remarkable for their rich flora of temperate aspect, have usually been referred to the Miocene, but it now seems certain that the greater part are really of Laramic or Eocene age. This seems evident from their fossil flora as described by Heer and others.

2. *Mackenzie River District and Arctic Coast.*—According to Richardson and other observers, the valley of the Mackenzie River from Arthabasca Lake to the Arctic Sea is occupied principally by Palæozoic rocks of Silurian and Devonian age with probably a narrow belt of Cambrian in the hills west of the mouth of the river. The remainder of the river valley, more especially in its southern part and toward the Peace River, is occupied with Cretaceous and Laramie beds, the latter affording in some places abundance of characteristic plants which have been described by Heer.

On the coast, west of Mackenzie River, the Cretaceous and Laramie beds extend as far as Cape Parry. From this place to Coronation Gulf, at the mouth of the Coppermine River, quartzite, slate and limestone, supposed to be of Cambrian age, predominate. In the islands to the north and east are great breadths of Siluro-Cambrian and Silurian limestones, and the peninsula of Boothia consists of an axis of Laurentian with Siluro-Cambrian and Silurian rocks which also line the great Laurentian mass between the Gulf of Boothia and Baffin's Bay.

3. *Hudson Bay.*—The basin of Hudson Bay, as described by Rae and Bell, is surrounded by Laurentian rocks bordered by Huronian and Palæozoic rocks, including fossiliferous Silurian limestones. On the west side these rocks extend northward to Cape Esquimaux and occur also on the south side of Hudson Bay, the wide tract of Laurentian rocks which extends to the Labrador coast, forming the shore of the bay, except toward the south, where rocks of Cambrian age are reported. They consist of limestones, sandstones and quartzites, shales and ironstones, associated with basalts and amygdaloids. At the south end of James' Bay are Devonian rocks from which Dr. Bell has obtained numerous species of fossils determined by Mr. Whiteaves.* At the south end of Hudson's Bay are also Pleistocene beds with lignite and marine fossils, *Saxicava*

* Survey of Canada, 1877-8.

rugosa, Macoma calcarea, Mya truncata. Bones of Mastodon have also been found.

4. *The Arctic Archipelago and coast of Greenland.*—In the appendix to McClintock's Voyage, Prof. Haughton has described this region, and arranges the formations represented as follows :—

1. Granitic and Granitoid rocks, probably in great part Laurentian gneiss, &c.

2. Silurian, consisting of Red Sandstone overlaid by fossiliferous limestones.

3. Carboniferous rocks, including limestone beds (Horton Series) with plants and marine limestones.

4. Lias or Mesozoic rocks holding *Ammonites, Monotis,* &c., and a Saurian, *Arctosaurus Osborni.*

5. Cretaceous and Tertiary, apparently of the same age with those on the coast of Greenland, and holding fossil plants.

6. Superficial deposits, which seem to be for the most part sands and clays with marine fossils.

The Carboniferous beds of the Arctic islands are thus described by Haughton :—"The Upper Silurian limestones, already described, are succeeded by a most remarkable series of close-grained white sandstones, containing numerous beds of highly bituminous coal, and but few marine fossils. In fact, the only fossil shell found in these beds, so far as I know, in any part of the Arctic Archipelago, is a species of ribbed *Atrypa,* which I believe to be identical with the *Atrypa fallax* of the Carboniferous slate of Ireland. These sandstone beds are succeeded by a series of blue limestone beds, containing an abundance of the marine shells, commonly found in all parts of the world where the Carboniferous deposits are at all developed. The line of junction of these deposits with the Silurian on which they rest is N.E. to E.N.E. (true.). Like the former, they occur in low, flat beds, sometimes rising into cliffs, but never reaching the elevation attained by the Silurian rocks in Lancaster Sound.

"Coal, sandstone, clay-ironstone and brown hæmatite were found along a line stretching E. N. E. from Baring Island, through the south of Melville Island, Byam-Martin Island, and the whole of Bathurst Island. Carboniferous limestone, with characteristic fossils, was found along the north coast of Bathurst Island, and at Hillock Point, Melville Island."

From a comparison of different coal exposures noted by McClintock, McClure, Austen, Belcher and Parry, in the Parry Islands, Prof. Haughton has laid down the approximate outcrops of some of the coal beds. These he finds to agree remarkably well with the trend of the boundary of the formation drawn from totally different data. In connection with this place it is noted that the Carboniferous sandstones underlie the limestones, and that "it is highly probable that the coal beds of Melville Island are very low down in the series, and do not correspond in geological position with the coal beds of Europe."

The following account of the Cretaceous and Tertiary of Greenland is condensed from Heer:—

I. CRETACEOUS.

1. The *Komé* series, of black shales resting on the Laurentian gneiss. These beds are found at various other localities, but the name above given is that by which they are generally known. Their flora is limited to ferns, cycads, conifers, and a few endogens, with only *Populus primava* to represent the dicotyledons. These beds are regarded as Lower Cretaceous (Urgonian,) but the animal fossils would seem to give them a rather higher position. They may be regarded as equivalent to the Kootanie and Queen Charlotte beds in Canada, and the Potomac series in Virginia.

2. The *Atané* series. These also are black shales with dark-coloured sandstones. They are best exposed at Upernavik and Waigat. Here dicotyledonous leaves abound, amounting to ninety species, or more than half the whole number of species found. The fossil plants resemble those of the Dakota series of the United States and the Dunvegan series of Canada, and the animal fossils indicate the horizon of the Fort Pierre or its lower part. They may be regarded as representing the lower part of the Upper Cretaceous. The genera *Populus*, *Myrica*, *Quercus*, *Ficus*, *Platanus*, *Sassafras*, *Laurus*, *Magnolia*, and *Liriodendron* are among those represented in these beds, and the peculiar genera *Macclintockia* and *Credneria* are characteristic. The genus *Pinus* is represented by five species, *Sequoia* by five, and *Salisburia* by two, with three of the allied genus *Baiera*. There are many ferns and cycads.

3. The *Patoot* series. These are yellow and red shales, which seem to owe their colour to the spontaneous combustion of pyritous lignite, in the manner observed on the South Saskatchewan and the Mackenzie rivers. Their age is probably about that of the Fox-Hill group or Senonian, and the Upper Cretaceous of Vancouver Island, and they afford a large proportion of dicotyledonous leaves. The genera of dicotyledons are not dissimilar from those of Atané, but we now recognise *Betula* and *Alnus*, *Comptonia*, *Planera*, *Sapotacites*, *Fraxinus*, *Viburnum*, *Cornus*, *Acer*, *Celastrus*, *Paliurus*, *Ceanothus*, *Zizyphus*, and *Cratægus* as new genera of modern aspect.

On the whole there have been found in all these beds 335 species, belonging to 60 families, of which 36 are dicotyledonous, and represent all the leading types of arborescent dicotyledons of the temperate latitudes. The flora is a warm temperate one, with some remarkable mixtures of sub-tropical forms, among which perhaps the most remarkable are *Kaidocarpum* referred to the *Pandaneæ*, and such exogens as *Ficus* and *Cinnamomum*.

II. TERTIARY.

4. The *Unartok* series. This is believed to be Eocene. It consists of sandstone, which appears on the shores of Disco Island, and possibly at some other places on the coast. The beds rest directly and apparently conformably on the Upper Cretaceous, and have afforded only eleven species of plants. *Magnolia* is represented by two species,

Laurus by two, *Platanus* by two, and one of these said to be identical with a species found by Lesquereux in the Laramie,* *Viburnum, Juglans, Quercus,* each by one species ; the ubiquitous Sequoias by *S. Langsdorffi.* This is pretty clearly a Lower Laramie flora.

5. The *Atanekerdluk* series, consisting of shaly beds, with limestone intercalated between great sheets of basalt, much like the Eocene of Antrim and the Hebrides. These beds have yielded 187 species, principally in bands and concretions of *siderite,* and often in a good state of preservation.   They are referred to the Lower Miocene, but, as explained in the text, the flora is more nearly akin to that of the Eocene of Europe and the Laramie of America.   The animal fossils are chiefly fresh-water shells. *Onoclea sensibilis,* several conifers, as *Taxites Olriki, Taxodium distichum. Glyptostrobus Europæus,* and *Sequoia Langsdorffi,* and 42 of the dicotyledons are recognized as found also in American localities.   Of these, a large proportion of the more common species occur in the Upper Laramie of the Mackenzie River and elsewhere in north-west Canada, and in the western United States.   It is quite likely also that several species regarded as distinct may prove to be identical.

It would seem that throughout the whole thickness of these tertiary beds the flora is similar, so that it is probable it belongs altogether to the Eocene rather than to the Miocene.

No indication has been observed of any period of cold intervening between the Lower Cretaceous and the top of the Tertiary deposits, so that, in all the vast period which these formations represent, the climate of Greenland would seem to have been temperate.   There is, however, as is the case farther south, evidence of a gradual diminution of temperature.   In the Lower Cretaceous the probable mean annual temperature in latitude 71° north is stated as 21' to 22° centigrade, while in the early Tertiary it is estimated at 12° centigrade.   Such temperatures, ranging from 71° to 53° of Fahrenheit, represent a marvellously warm climate for so high a latitude.   In point of fact, however, the evidence of warm climates in the arctic regions, in the Palæozoic as well as in the Mesozoic and early Tertiary, should perhaps lead us to conclude that, relatively to the whole of geological time, the present arctic climate is unusually severe, and that a temperate climate in the arctic regions has throughout geological time been the rule rather than the exception.

It is interesting to note that the driftage of erratics in the Arctic basin seems to have been northward.   The following extract refers to this. †—

"Along the Arctic coast, and among the islands of the archipelago, there is a considerable volume of evidence to show that the main direction of movement of erratics was *northward.*   Thus, boulders of granite, supposed by Prof. Haughton to be derived from North Somerset, are found 100 miles to the north-eastward, (Appendix to McClintock's Voyage, p. 374,) and pebbles of granite, identical with that of Granite Point, also in North Somerset, occur 135 knots to the north-west.   (Op. cit. p. 376.)   The east side of King William Land is also said to be strewn with boulders like the gneiss of Montreal Island to the southward. Prof. Haughton shows the direction and distance of travel of some of

---

* *Viburnum marginatum* of Lesquereux.

† Dr. Dawson's Notes, p. 57.

these fragments by arrows on his geological map of the Arctic archipelago, and reverts to the same subject on pages 393 and 394, pointing out the general northward movement of ice indicated, and referring the carriage of the boulders to floating ice of the glacial period.

"Near Princess-Royal Island, in Prince-of-Wales Strait, and also on the coast of Prince-of-Wales Island, the copper said to be picked up in large masses by the Eskimo, (De Rance, Nature, Vol. xi. p. 492,) may be supposed to be derived from the Cambrian rocks of the Coppermine river region to the south, as it is not probable that it occurs in place anywhere in the region of horizontal limestone where it is found.

"Dr. Armstrong, previously quoted, notes the occurrence of granitic and other crystalline rocks, not only on the south shore of Baring Land, but also on the hills inland. These, from what is now known of the region, can scarcely be supposed to have come from elsewhere than the continental land to the southward.

"In an account of the scientific results of the 'Polaris' expedition, (Nature, vol. ix,) it is stated of the west coast of Smith's Sound, north of the Humboldt glacier, that 'wherever the locality was favorable,' the land is covered by drift, sometimes containing very characteristic lithological specimens, the identification of which with rocks of South Greenland, was a very easily accomplished task. For instance, garnets of unusually large size were found in lat. 81° 30′, having marked mineralogical characters by which the identity with some garnets from Tiskernaces was established. Drawing a conclusion from such observations, it became evident that the main line of the drift, indicating the direction of its motion, runs from south to north. It should be stated, however, that Dr. Bessels, who accompanied the 'Polaris' expedition, regards these erratics as certainly not transported by glaciers, but by floating ice, and as showing that the current of Davis Strait was formerly to the north and not to the south, as at present. (Bull. Soc. Geog., Paris, vol. ix., 1885, p. 297.)

"It may further be mentioned as bearing on the general question here referred to, that Dr. Bell has found evidence of a northward or northeastward movement of glacier-ice in the northern part of Hudson Bay, (Annual Report Geol. Survey of Canada, 1875, p. 14, D.D,) with distinct indications of eastward glaciation throughout Hudson Strait. (Report of Progress, Geol. Survey of Canada, 1882-84, p. 36, D.D.)

The facts so far developed in this northern part of the continent and in the Arctic islands, thus point to a movement of ice outward in all directions from the great Laurentian axis or plateau, which extends from Labrador round the southern extremity of Hudson's Bay to the Arctic Sea, rather than to any general flow of ice from the vicinity of the geographical pole southward.

VI. THE TERRANOVAN OR NEWFOUNDLAND REGION.

A large part of the Island of Newfoundland is occupied with Laurentian rocks. The western booundary of these extends south-west throughout the length of the island, from near Hare Bay at the northern extremity of the island to Cape Ray at its south-west end. The eastern boundary of the Laurentian forms the west coast of White Bay, and south-east of this is broken by belts of newer strata, beyond which it forms a sinuous line from Cape Freels on the east coast, to Fortune Bay. The extreme breadth of this Laurentian district is therefore about 150 miles, and the general strike of its beds seems to be N. E. and S.W. Its continuity is, however, interrupted by two areas of Palæozoic rocks resting on it. One of these is a great tract of Cambrian and Siluro-Cambrian rocks presenting a broad front to the Atlantic at Notre Dame Bay, and forking into three branches before reaching the south coast. It includes great breadths of serpentine, associated with chloritic slates, diorites, etc., and also a long trough of Silurian rocks.

The second and narrower break is that stretching to the south-west from White Bay, along the Humber River to St. George's Bay, where it is united with the Palæozoic of the west coast. It contains Siluro-Cambrian, Silurian and Carboniferous rocks.

To the westward of the Laurentian nucleus, the Gulf of St. Lawrence coast is skirted with formations reaching in age from the Potsdam to the Carboniferous, and which, in their northern part, seem to be faulted against the Laurentian.

To the eastward of the Laurentian nucleus the south-eastern portion of the Island, including the peninsula of Avalon, is occupied, with the exception of two considerable patches of Laurentian, by rocks referred by Murray to the Huronian and Lower and Middle Cambrian.

These eastern rocks partake of the marginal and Atlantic character of those of Nova Scotia and New Brunswick, of which the eastern part of Newfoundland may be regarded as a north-eastern continuation. The Palæozoic formations on the west coast, on the other hand, constitute the eastern margin of the great undisturbed area of the Gulf of St. Lawrence.

The Huronian rocks of Newfoundland consist of quartzites, diorites, slates and slate conglomerates, resembling those of the typical region in Georgian Bay. Their upper beds have, however, afforded the doubtful fossils known as *Aspidella* and *Arenicolites spiralis*.

Above these are red and green sandstones and conglomerates known as the Signal Hill series. They correspond in mineral character and geological relations to the Kewenian of the west, and the Basal Cambrian of Matthew, in New Brunswick.

Next in succession are the Cambrian beds, consisting largely of shales, sometimes slaty, and sandstones, with some limestone. They appear to include the lower members of the Cambrian system as developed on the Atlantic coast. In a recent paper,* Walcott gives the section seen on Manuel's Brook, Conception Bay, as follows :

a. Archæan Gneisses.   Feet.

1. Conglomerate resting uncomformably upon *a*........   35
2. Sandstone, shale and impure limestone, with *Olenellus Broggeri*
       sixteen species of the Olenellus fauna..   25
3. Greenish argillaceous shale.... 40
4. Red   do.   do.   4
5. Limestone..........   2
6. Greenish argillaceous shales with an abundant *Paradoxides* fauna
       at summit... 270
7. Dark argillaceous shales, *Paradoxides*, *Microdiscus punctatus*,
       Agraulus, *Conocoryphe*, etc., etc., near base.................. 295
8. Alternating bands of shale and sandstone, with *Orthis* in great
       abundance 400
                                                                   ────
               Total,..... 1071

Dip of Strata, 12° to 15° N. All strata unaltered and undisturbed.

This section includes the Lower and Middle Cambrian, and, according to Walcott, proves that the *Olenellus* zone is below that of *Paradoxides*.

The Upper Cambrian or *Olenus* and *Dikellocephalus* zone does not seem to be recognized in southern Newfoundland, but occurs on the strait of Belle Isle on the north, where it is continuous with the west coast rocks of the interior type.

Rocks equivalent to the Quebec group of Sir W. E. Logan, occupy a considerable area in northern and western Newfoundland, and consist of shales, sandstone and conglomerate, with beds of limestone, containing fossils characteristic of the Calciferous and Chazy horizons. Like those on the Lower St. Lawrence, they are often thrown into sharp folds. As on the east side of the island, large areas of serpentine, accompanied with altered slates, are associated with these rocks. The valuable copper deposits of Tilt Cove Section are associated with these

─────────
* Nature, Oct. 1888.

serpentines. In the narrow peninsula of Long Point, at the west side of Port-à-Port Bay, there are limestones which contain fossils referable to the Trenton period, and which may indicate the outer margin of an extensive area of such limestones under the waters of the Gulf of St. Lawrence, and resembling this formation as it occurs in the interior plateau of North America and in the valley of the river St. Lawrence.

The Silurian beds are represented chiefly in the troughs running southwest from White Bay and the Bay of Exploits. They consist of conglomerates below, overlaid by shales of various qualities, and beds of sandstone, but with only a meagre development of limestone, in this resembling the Silurian of Nova Scotia. In the Bay of Exploits region, they are traversed by trap dykes and in places overlaid by volcanic breccia.

The fossils noted are *Favosites Gothlandica, Heliolites, Zaphrentis bellistriata, Strophomena rhomboidalis, Atrypa reticularis, Stricklandinia lens,* etc.

On the north side of White Bay, and at its upper extremity, are small areas of sandstones and conglomerates, in which fossil plants have been found. These appear to be of Erian or Devonian age and the beds themselves resemble the Gaspé sandstones. They occupy, however, only very small areas so far as known.

The newest formation in Newfoundland, except the Pleistocene, is the Carboniferous, which is limited to the trough extending from St. George's Bay on the west coast to White Bay on the east. It consists of two detached areas, the larger fronting on St. George's Bay, and the other on the head waters of the Humber River, not far from the head of White Bay. In both areas beds occur representing the Lower Carboniferous limestone, the Millstone Grit and the coal formation, or the lower part of it. As in Nova Scotia, beds of gypsum occur in connection with the limestone on St. George's Bay.

Murray gives the following general section of the formation as developed on St. George's Bay, in descending order :

Green and red sandstones with brown, drab and black shales and clays; coal and fossil plants....	1000
Brown and reddish sandstones and conglomerates and shales; fossil plants and thin layers of coal................	2000
Variegated marls and red, green and brown sandstones, with beds of bluish and gray limestone, rich in marine fossils.....	2000
Beds of gypsum with marls, sandstones and shales, and dark-colored limestones and shales	150
Coarse conglomerate with boulders and pebbles of Laurentian and Silurian rocks, lenticular beds of sandstone and shale........	1300
	6450 feet.

The upper member of the above corresponds to the lower part of the productive coal formation in Nova Scotia. In Newfoundland the most important bed of coal yet found in it is four feet thick. The second member is the equivalent of the millstone grit, the third and fourth of the Windsor series of Nova Scotia, and the last of the Horton series, in that form which it assumes on the south side of the Cobequids in Nova Scotia and in the Bonaventure series in Quebec and New Brunswick.

In specimens collected by Dr. R. Bell, at St. George's Bay, I have recognized the following species, most of them found also in Nova Scotia :—*Serpulites annulatus*, Dawson; *Conularia planicostata*, Dawson; *Aviculopecten Dcbertianus*, Dawson; *Bakevellia antiqua*, Mont; *Pteronites Gayensis* (var. *ornatus*,) Dawson; *Cypricardia*, sp. *Terebratula sacculus*, Martin; *Spirifera glabra*, Martin; *Productus semircticulatus*, Martin; *P. Cora*, D'Orbigny; *Streptorhynchus crenistria*, Phillips. There is also a new Serpulites, (*S. Murraii*, Dawson,) and a beautiful little Gastropod, which I have named *Macrocheilus Terranovicus*.

In a small collection of fossil plants forwarded by Mr. Murray there are two species of *Lepidodendron*, one of them apparently new, ferns of the genus *Sphenopteris* and fossil wood of the genus *Dadoxylon*.

The Pleistocene deposits are represented in Newfoundland by boulder clay and by clays holding marine shells of recent species, and surrounded by sand and gravel, as in most other parts of eastern America. So far as known the direction of drift has on the low grounds been from the northeast, and there seems also evidence of local drift and striation from the central highlands down the valleys toward the coast. These directions mark different stages of the great Pleistocene submergence, in which Newfoundland seems to have shared like other parts of eastern America. In Newfoundland, according to Richardson and Hinde, there is evidence of submergence to the amount of 1200 feet, and hills at the height of 2500 feet show evidence of ice-action.*

The economic resources of Newfoundland, as described by Murray and Howley, include copper ores (Tilt Cove, etc.), galena (Port-à-Port,) iron ores, pyrites and ochres, gypsum and coal. Gold occurs in quartz veins near Brigus in Conception Bay, in rocks regarded as Huronian. For more full accounts of the economic minerals of the island, reference may be had to the Reports of Messrs. Murray and Howley, of the Geological Survey of Newfoundland.

---

* For the details see the author's Notes on the Post-Pliocene, 1872.

APPENDIX.

The following descriptions are intended to afford the student convenient means of reference to the characters of the more important minerals referred to in the chapter on Lithology.

1. QUARTZ.

As familiar examples, Flint and Rock Crystal may be taken. The former occurring in concretions in chalk and other calcareous rocks, was probably one of the first mineral substances used by man ; being the material of the flint implements of the "stone age." As quartz is the most common of minerals, and occurs in most silicious rocks, it may serve as a typical mineral whereby to illustrate the terms used in other cases.

Composition.—Quartz when pure is *silica*, a compound of the elements *silicon* and *oxygen.* The former is an element not unlike carbon or charcoal in many of its properties; the latter a gas and the most important ingredient of the atmosphere. Silica is thus an *oxide of silicon*, and containing two combining proportions of Oxygen to one of Silicon, its chemical name is *silicon dioxide.*

Crystallization.—Its usual form is a six-sided prism, terminated by a six-sided pyramid. It thus belongs to the *Hexagonal* system of crystallization. When mineral substances solidify from the state of vapour, from solution in water, or from a state of fusion, their particles tend to arrange themselves along certain lines or axes, and thus to produce crystals of definite geometrical forms. The law in the case of quartz is, that its particles arrange themselves along three horizontal axis, or lines of attraction, at angles of sixty degrees with each other, and along a fourth axis at right angles to the other three. The six-sided plates and six-rayed stars of snow are formed on the same principles.

Perfect crystals of quartz are found lining *geodes* or cavities in rocks, also the sides of fissures and veins, and sometimes imbedded in the substance of rocks. Small crystals confusedly aggregated and imperfect, owing to pressure, give *granular varieties.* Crystals so small that they cannot be discerned by the naked eye give *cryptocrystalline varieties.*

Its *Hardness* is 7, measured by a scale in which talc is 1 and diamond 10. The hardness of quartz is sufficient to enable it to scratch glass, to resist the action of steel, and to feel gritty in the teeth.

Its *Specific Gravity* is 2.5 to 2.8, measured by a scale in which water is the unit. It is thus two and a half times heavier than water. Quartz being one of the most common minerals, and entering very largely into the composition of rocks, in which also it is associated with many other substances not very different in specific gravity, it follows that its specific gravity is about that of most ordinary rocks, all of which are thus sufficiently heavy to sink readily in water, but when immersed in water lose between one half and one third of their weight.

Optical Characters.—Quartz is sometimes *colourless,* but becomes coloured by mixture with other substances, especially oxides of iron. The protoxide (ferrous oxide) gives dull green and blackish colours—the peroxide (ferric oxide) red colours, and the hydrous peroxide yellow and brown colours. The *lustre* of quartz is, with reference to its *kind, vitreous* or that of broken glass; with reference to its *degree,* it varies from splendent, the lustre of perfect crystalline faces, to dull or lustreless. The vitreous lustre is a good character whereby to distinguish the mineral. The pure and crystalline varieties are *transparent ;* the crypto-crystalline and coarse varieties *translucent* to *opaque.*

Quartz is *brittle,* and its fracture *conchoidal* in the pure varieties. It is *infusible* and *insoluble* in water and ordinary acids ; but may be fused or dissolved in water, when combined with alkalis, as potash or soda.

VARIETIES OF QUARTZ.

Quartz presents many varieties, which may be arranged under the heads of (*a*) Crystalline or vitreous, (*b*) Chalcedonic and (*c*) Jaspery.

(a.) *Vitreous Varieties.*

Rock Crystal.—Transparent and colourless, often in the definite crystalline form. Used for lenses, for ornamental purposes and to form imitation gems or doublets.

Amethyst.—Purple and violet varieties, coloured by a minute quantity of Manganese, or perhaps in some cases by iron and soda.

Rose Quartz.—A more delicately tinted pink variety.

Yellow and Smoky Quartz.—Called Cairngorm or false Topaz, of smoky or rich yellowish and brownish hues; coloured by titanic acid, or by organic matter.

Cat's Eye.—Transparent quartz, containing fibres of asbestos, which give it a lustre resembling that of satin.

Aventurine is translucent quartz spangled with brilliant scales of yellow mica.

The glassy varieties of quartz pass into common *milky quartz*.

(b.) *Chalcedonic Varieties.*

Chalcedony is the general name for colourless varieties having a glistening and somewhat waxy lustre.

Carnelian is a flesh coloured or red chalcedony coloured by iron. A deeper red variety is called *sard*.

Agate is chalcedony with bands or spots of different textures and colours. When these are in parallel layers it is called *onyx*. Some of the layers being absorbent, colourless agates of this kind can be artificially coloured. When some of the layers are of carnelian or sard it is called *sardonyx*. These banded varieties are the material of cameos. When translucent chalcedony is penetrated with moss-like or dendritic filaments of oxide of iron or manganese it is *moss agate* or *Mocha stone*.

Chrysoprase is a pale green variety coloured by oxide of nickel. Green varieties coloured merely by iron are common *Prase*.

Flint and *Chert* are names for coarse chalcedonic varieties, usually impure and of dull colours. A cellular variety is *Buhr-stone*, used for millstones.

Agates are produced by the deposition of chalcedony in the cavities of rocks, usually those of volcanic or igneous origin. When this process is slow or intermittent, bands of various textures and colours are formed in succession. Flint and Chert are formed by the slow collection of silicious particles around centres by concretionary action. Hence they often have fossil sponges, etc., in their interior. Wood imbedded in rocks is often fossilized by silica, or silicified, so as to resemble agate.

(c.) *Jaspery Varieties.*

Jasper includes those varieties which are opaque and more or less deeply coloured, usually by oxide of iron. Red jasper is one of the most common varieties; a brown coloured or banded jasper is called *Egyptian jasper;* green and red or green and yellow banded varieties, are *riband jasper;* a green variety with bright red spots is *bloodstone* or *heliotrope.*

Quartz occurs very largely in the earth's crust, as sand, sandstone and quartz-rock or quartzite, and also as a constituent of many compound rocks.

2. FELSPAR.

There are several species of felspar; but we may take as an example the most abundant and most important species, *orthoclase* or common felspar.

In *Chemical Composition* it is a silicate of alumina and potash. It is not acted on by acids and is fusible with difficulty.

Its *Crystalline form* is monoclinic; its particles being arranged in accordance with three axes of crystallization—the two horizontal ones at right angles to each other, the third at an angle of 63° 53' to plane of the others. Its two principal cleavages are at right angles, while the corresponding cleavages in the other felspars are never at right angles. These cleavage faces aid in distinguishing it from quartz.

Its *Hardness* is six, being thus next to quartz in the scale of hardness. Though scratched by quartz it is hard enough to scratch glass, but feebly.

Its *Specific Gravity* is 2.5 to 2.6.

Its *Lustre* is vitreous, but with a tendency to pearly on cleavage surfaces. Its more common colors are white, red and grey.

Kaolin, or the finest China clay, proceeds from the decomposition of felspar; the potash and part of the silica being dissolved by rain water and leaving a hydrous silicate of alumina in a fine state of division.

Of the other Felspars the most important are *Albite,* soda felspar, in which soda replaces the potash of the orthoclase, *Anorthite,* or lime felspar, and *Oligoclase* and *Labradorite,* in which both soda and lime are present. Albite sometimes presents a beautiful pearly opalescence upon its cleavage faces, and Labradorite is remarkable for the splendid play of colours observed in some specimens. Labradorite, Anorthite and Oligoclase are basic, or have an excess of base relatively to their silica. All beside Orthoclase are triclinic.

The felspars are extremely important in geology as constituents of the silicious crystalline rocks, as granite, syenite, gneiss, dolerite, etc.

They enter very largely into the composition of the lavas of volcanoes; those called *Trachytic Lavas* or *Trachytes,* consisting principally of felspar.

3. LEUCITE.

This mineral, like orthoclase, is a silicate of alumina and potash; the constituents, however, being in different proportion. Crystallization isometric, the crystals often being very perfect trapezohedrons, with twenty-four similar faces.

H.—5.5 to 6. Gr.—2.45 to 2.5.

Colour, white to gray; surface usually dull; infusible. Decomposed by hydrochloric acid, without gelatinization. Not known to occur in Canada, but an abundant constituent of certain volcanic rocks in Italy and the United States.

4. NEPHELINE.

Nepheline is a mineral which sometimes takes the place of felspar in crystalline rocks. It is a silicate of alumina, soda and potash, and crystallizes in hexagonal forms. The color is commonly white or gray, and the lustre vitreous to greasy.

H.—5.5 to 6. Gr.—2.5 to 2.6.

Nepheline is decomposed by hydrochloric acid, with separation of gelatinous silica. It is a prominent constituent of nepheline-basalt, nepheline-syenite, phonolite and other rocks.

5. MICA.

Of this also there are several species: Common mica or *Muscovite* is the most important.

It is a very complex silicate, containing silica, alumina, potash, iron, magnesia, lime and soda.

Its crystals are inclined rhombic and six-sided prisms (monoclinic). The angles of the rhombic prisms are 120° and 60°. It is remarkable for its very perfect cleavage parallel to the base of the prism. In this direction it may be split into extremely thin laminæ, which are flexible and elastic. When crystallized in small radiating plates it is called *Plumose Mica.*

H.—2.0 to 2.5. Gr.—2.75 to 3.1.

Its lustre on the faces of the cleavage planes is metallic pearly, and its colors range from silvery white to greenish, yellow and black. They are due to oxides of iron.

Along with quartz it forms mica-schist, and in a very fine state of division it is largely concerned in giving cleavage to roofing slate. It also gives a flaggy character to sandstone. It general, when scales of mica are arranged in parallel layers in rocks, they give to these more or less of their own fissile character.

Biotite, a mica containing much magnesia and iron, and of a dark color, is next in importance to Muscovite.

Phlogophite is a mica containing a large proportion of magnesia, and commonly less iron than Biotite. It is often of a curious brownish-red color. It is one of the most common minerals in the Apatite-bearing veins of Canada, as well as in some of the Laurentian limestones.

6. PYROXENE.

The name of this mineral, implying that it is a "stranger to fire," is a reminiscence of the old controversies as to the origin of rocks from water or heat, and is curiously contrary to the fact that Pyroxene is one of the largest constituents of volcanic rocks.

Composition.—Silica, lime, magnesia and iron. Some of the varieties have much more iron than others.

Crystalline form, monoclinic or like that of orthoclase, but the angles different, the inclination of the principal axis being 73° 59', and the acute angle of the rhombic prism 87° 5', so that it is nearly square. It occurs also in granular and fibrous forms. Its cleavage is not perfect, but may be obtained parallel to the faces and bases of the prisms.

H.—5 to 6. Gr.—3.2 to 3.5.

It is thus almost as hard as felspar, and somewhat heavier than that mineral or quartz, so that rocks containing much pyroxene are usually somewhat heavy.

It ranges in colour from white, through different shades of grey and green to black, the chief colouring constituent being ferrous oxide. Its lustre is vitreous inclining to resinous, and in some varieties it becomes pearly.

Varieties.

These are very numerous and have received different names. We shall notice only a few of some geological importance.

Augite or common Pyroxene. This is of dark colour, usually black; and is the form in which the mineral most commonly occurs as an ingredient in rocks.

Sahlite and *Malacolite* are light green and white varieties, also occurring sometimes as considerable ingredients of rocks.

Diallage is a variety with a very distinct cleavage, and strong metallic pearly lustre on the surfaces.

7. HORNBLENDE.

This is a mineral closely allied to Pyroxene. Its ordinary varieties, however, contain more magnesia and less lime than the latter.

Crystallization monoclinic, but its rhombic prism is much flatter than that of Pyroxene, its obtuse angle (corresponding in position to the acute angle in the case of Pyroxene) being 124° 30′, and it has a distinct cleavage parallel to the sides of the prism. It thus forms flat blade-like crystals, and those being often long and slender, it assumes fibrous forms.

H.—5 to 6.   Gr.—2.9 to 3.4.

Its range of colour is similar to that of the last species.

Varieties.

Common Hornblende or Amphibole includes the dark and more massive varieties.

Actinolite is green, and columnar or fibrous.

Tremolite is white or gray, and finely fibrous.

Asbestos includes the finest fibrous varieties, which, from the slenderness and flexibility of the fibres, may be woven into fabrics which have become celebrated as incombustible cloths.

Mountain Wood, Mountain Cork and *Mountain Leather* are fibrous and lamellar varieties resembling the substances whose names they bear.

8. OLIVINE.

This is a silicate of magnesia and ferrous oxide, and crystallizes in the ortho-rhombic system. It is commonly met with in rounded grains imbedded in crystalline rocks rather than as well defined crystals.

H.—6 to 7.   Gr.—3.3 to 3.6.

Colour usually olive green, sometimes yellow, brownish, reddish. Lustre, vitreous. Decomposed by hydrochloric acid with separation of gelatinous silica. Olivine is frequently altered to serpentine. It is a constituent of many crystalline rocks, some of which are almost entirely composed of it. A rock of this kind (dunite) occurs at Mount Albert, in the province of Quebec.

9. GARNET.

This is a mineral which varies greatly in composition, and has accordingly been divided into a number of subspecies. It is a silicate of different sesquioxides and protoxides (sesquioxides of aluminum, iron or chromium, and protoxides of iron, calcium, magnesium or manganese.)

The crystallization is isometric; rhombic, dodecahedrons and trapezohedrons being the common forms. In many cases garnet occurs imbedded in rocks in rounded or irregular grains.

H.—6.5 to 7.5. Gr.—3.1 to 4.3.

Commonly red in colour, but also green, yellow, black, etc.; Lustre, vitreous; insoluble in hydrochloric acid and most varieties infusible. Garnet is frequently found in such rocks as gneiss and mica-schist.

10. TALC.

Is a silicate of magnesia, with water. It is thus an example of a hydrous silicate.

Crystallization ortho-rhombic, and usually occurring in foliated or cleavable masses, the cleavage being similar to that of mica. It also occurs massive or crypto-crystalline,

H.—1. Gr.—2.5 to 2.8.

The low hardness of Talc affords a ready means of distinguishing it from other foliated minerals. It has also a soapy or unctuous feel, and its laminæ are not elastic.

Its colour is usually light green, though sometimes a silvery white. Its lustre is pearly.

Soapstone and *Potstone* are compact or confusedly crystalline varieties, used for firestones, for furnaces, or vessels required to stand the fire.

French Chalk is a variety used for marking.

Talc is an ingredient in Talc schists, to which it communicates its own foliated character.

Meerschaum is closely allied to Talc, but has a larger proportion of water.

11. SERPENTINE.

This is a silicate of magnesia with water, the latter in larger quantity

than in talc.   It usually occurs massive, and sometimes fibrous.   It sometimes constitutes considerable rock masses.

H.—2.5 to 4.   Gr.—2.5 to 2.6.

Its colour is usually green, and its lustre somewhat resinous or waxy.

Precious Serpentine includes varieties of a rich green colour and translucent.   *Common Serpentine* includes the more dull-coloured and opaque
varieties.   *Picrolite* and *Chrysotile* are fibrous varieties ; the latter often
called Asbestos, and serves the same purposes with that mineral.   *Ophiolite*
or *Verde Antique Marble*, consists of a mixture of Serpentine and Calcite,
and is usually of green and white colours.

12. CHLORITE.

This represents a group of several species or sub-species.   Chlorite may
be regarded as a hydrous silicate of alumina, magnesia and iron.   It
occurs in foliated masses and flat crystals, of a greenish colour and slightly
pearly lustre.   It is harder than talc, and its laminæ are not elastic.   It
is the leading ingredient of chlorite schists.

13. CALCITE.

Is Calcium carbonate, or common Limestone.   Its effervescence with
acids, owing to the disengagement of gaseous carbonic acid, is one of
the ready ways of distinguishing it.   Its inferior hardness, enabling it to
be easily scratched with a knife, aids in distinguishing it from quartz,
felspar and other hard silicious minerals.

Crystallization hexagonal.   It occurs in many forms belonging to this
system ; especially the six-sided prism, the rhombohedron and the
scalenohedron.   It has very distinct cleavage parallel to the faces of the
rhombohedron.   It occurs also in granular, fibrous and crypto-crystalline
states, as well as in earthy conditions.

H.—3.   Sp. Gr.—2.5 to 2.8.

It is colourless, but is often coloured by other substances, especially
oxides of iron and carbonaceous matter.   Its lustre is vitreous, inclining
to pearly on the cleavage faces.   It varies from transparent to opaque.
The transparent varieties known as *Iceland Spar* possess double refraction.

Varieties.

CalcareousSpar includes the perfectly crystalline forms.

Satin Spar is a fibrous form occurring in veins, and having a silky
lustre.

Calc Sinter is a general name, which may include the imperfectly crystalline conditions occurring in *Stalactites* and *Stalagmite, Congealed water, Gibraltar Spar* and *Calcareous Tufa.* All these varieties are deposited from solution in water, aided by an excess of carbonic acid.

14. DOLOMITE.

This is calcium and magnesium carbonate. It effervesces less readily with acids than calcite. Its crystallization is rhombohedral like that of calcite, except that the angles of its rhombohedron are slightly different, and it is a little harder and heavier. It has also a more pearly lustre.

Dolomite occurs in nature in the same manner as calcite, but often contains ferrous carbonate, which causes it to assume a rusty colour in weathering.

15. GYPSUM.

Calcium Sulphate with a large proportion of water (about 20 per cent.) Its crystallization is monoclinic, and it has a very distinct cleavage, parallel to the larger faces of the rectangular prism. It is found in foliated, fibrous and granular crystallizations, and sometimes occurs in thick beds. Finer granular and translucent varieties are used for ornamental purposes, under the name of soft or gypseous alabaster. Its softness, enabling it to be scratched with the finger nail, and its pearly lustre, are distinguishing characters.

H.—1.5 to 2.   Gr.—2.31 to 2.33.

Its lustre is pearly upon the cleavage faces. It is colourless, but frequently stained red by ferric oxide, and sometimes black by carbonaceous matter.

Selenite is a lamellar variety of Gypsum. Fibrous varieties are used to imitate Cat's eye.

The readiness with which Gypsum parts with its water when heated, and resumes it, becoming solid or setting, when mixed with water, gives the substance important economical uses for casting, plastering and cements. It is the cheapest means of supplying sulphuric acid to the soil, and to manures, and thus is of some value in agriculture.

Anhydrite is anhydrous calcium sulphate. It is found with the previous species, from which it differs in its greater hardness and specific gravity, and its orthorhombic crystallization. It is sometimes used as an ornamental stone in the same manner as marble.

16. Apatite.

This is calcium phosphate with a small proportion of calcium chloride or fluoride, and is of great interest as representing the earthy part of the bones of animals.

Its crystallization is hexagonal, and its usual form is the hexagonal prism.

$$\text{H.—5.} \quad \text{Gr.—3. to 3.2.}$$

Its lustre is resinous, and its colour usually greenish.

In the crystalline state it occurs largely in veins and beds in the Laurentian formation in Canada. It is also found in concretionary masses, in beds of various geological ages, and is the principal constituent of the harder varieties of guano.

Calcium phosphate is an essential ingredient in soils, in which it is usually present in very small quantity, and it is rapidly removed by those crops which produce the greatest amount of animal food. This gives to it a very great importance in agriculture, and it is much sought for in every civilized country, and largely used as a means of improving the soil.

17. Fluor Spar or Fluorite.

This is Calcium Fluoride. Its crystalline form is isometric, and it often occurs in beautiful and regular cubes, with a cleavage parallel to the faces of the octahedron.

$$\text{H.—4.} \quad \text{Gr.—3. to 3.25.}$$

It is sometimes colourless, but more frequently of blue and purple colours, and in some cases green, red or yellow.

It frequently occurs in metallic veins, more especially with the ores of lead. It has been used as a flux in reducing metallic ores, hence its name Fluor.

18. Rock Salt.

Common Salt is sodium chloride. It crystallizes in the isometric system, usually in cubes.

$$\text{H.—2.} \quad \text{Gr.—2.1 to 2.25.}$$

It furnishes an excellent example of a soluble native salt. It occurs not only in great quantity in the sea and in salt lakes, but also in extensive beds in the crust of the earth, whence it is mined for use. These beds have probably been formed by the drying up of salt lakes, and of isolated portions of sea water, and the subsequent covering by sediment of the beds of salt thus formed. Copious salt springs often rise from such deposits.

19. MAGNETITE.

Is an oxide of iron intermediate between the monoxide and sesquioxide. Crystallization isometric, usually in octahedrons.

H.—5.5 to 6.5. Gr.—5.

Colour black, Lustre metallic. It occurs in Canada in large beds, in the Laurentian, and also in layers as iron sand, and is the most valuable of the ores of iron. It is attracted by the magnet, and it sometimes has itself magnetic polarity, constituting the natural loadstone. It is distinguished from the other species by its black powder or streak and its magnetic properties.

20. HEMATITE.

Also called specular iron, is sesquioxide of the metal. Its crystallization is hexagonal, and it often occurs in thin plates or scales, and also in fibrous forms.

H.—5.5 to 6.5. Gr.—4.5 to 5.3.

Its colour is black or steel grey in crystalline varieties, but its streak or powder is deep red. It is not usually attracted by the magnet.

Foliated varieties constitute *Micaceous Iron Ore*, compact or fibrous dull red varieties are called *Hematite*, and earthy varieties are *Red Ochre*. It is a very valuable ore of iron.

21. LIMONITE.

This is hydrous sesquioxide of iron. It occurs in fibrous and concretionary masses.

H.—5. to 5.5. Gr.—3.6 to 4.

Its colour is dark brown, and its streak or powder yellow. Compact and fibrous varieties are called *Brown Hematite*. Concretionary varieties found in modern deposits are *Bog Iron Ore*, and earthy varieties are *Yellow Ochre*. It is a valuable ore of iron.

22. PYRITE.

Is disulphide of iron. Crystallization isometric, usually in cubes and octahedrons.

H.—6. to 6.5. Gr.—4.8 to 5.

Colour, bronze yellow. It is a very common mineral, and is often mistaken for gold and for valuable metallic ores. When mixed with

metallic ores and with coal it is a troublesome impurity; but it is used as
source of sulphur and sulphuric acid, and of the ferrous sulphate.

23. COAL.

Coal essentially consists of compounds of carbon and hydrogen, with
variable amounts of oxygen, of nitrogen and of earthy matter. It presents many varieties, which shade into each other and differ much in
composition and physical properties. This results from the fact that it
is not a definite chemical compound, or crystallized mineral species, but
rather a product of the partial decomposition of vegetable matter buried
in the earth.

Its vegetable origin is proved by the remains of plants imbedded in it,
and often showing their structure distinctly under the microscope, and by
its resting on under-clays containing roots of trees, overlaid with shales
filled with impressions of plants. It is of different geological ages, but
the greater part was formed at a particular part of the earth's geological
history, known as the carboniferous period.

Its hardness varies from 1. to 2.5, and its sp. grav. from 1. to 1.8.
Its colour is black, or dark-brown, its powder either black or brown. Its
lustre is resinous or sub-metallic, and its fracture conchoidal or flat. It
usually presents a laminated structure, with layers of mineral charcoal, or
of vegetable debris, or of earthy matter between the laminæ, which often
consist principally of flattened trunks of which the coal has been made up.

The principal varieties are the following :—

Brown Coal is an imperfect coal found in the more modern formations.
It is often merely a consolidated peat, but when composed of flattened
trunks of trees it assumes the compact form of jet. It is intermediate in
composition between coal and wood. It contains from 47 to 70 per
cent. of carbon, and from 5 to 18 per cent. of hydrogen, the remainder
being oxygen and ashes. It is usually an inferior kind of fuel.

Bituminous Coal, or ordinary black coal, proceeds from a more perfect
carbonization of vegetable matter, and is the coal of the true Carboniferous
system. The coking varieties become soft when heated, and burn with
much flame. The non-coking varieties do not soften, and contain less
gaseous matter. Bituminous coal contains from 75 to 90 per cent. of
carbon, and from 3 to 6 per cent. of hydrogen, the remainder being
principally oxygen and ashes. The more bituminous varieties are used
for the production of gas.

Anthracite proceeds from the alteration of bituminous coals, and is sometimes of the nature of natural coke. It is harder and heavier than the bituminous coals, and contains from 85 to 92 per cent. of carbon and from 2 to 3 per cent. of hydrogen. It gives little or no flame in burning. In some coal deposits Anthracite passes by a further process of alteration into graphite or plumbago, which is, however, regarded as a distinct mineral species, owing to its very different physical properties.

24. BITUMEN.

Mineral oil and mineral pitch are mixtures of different hydro-carbons, differing from coal in their liquid, viscid or easily fusible character, and in being soluble in oil of turpentine and ether. Like coal, these substances are derived from the chemical change of vegetable matter buried in the crust of the earth; but they result chiefly from marine vegetation, or from that which has been buried and excluded from the air while still recent.

Petroleum, or mineral oil, includes the liquid or viscid varieties which flow from natural oil wells, or are obtained by boring into the beds of rock containing this substance in their pores and fissures. It has been known and used from the most ancient times, but has recently acquired greater importance from the abundance of it obtained by boring, and the means discovered for its purification. Petroleum often contains more than 12 per cent. of hydrogen.

Asphaltum includes the solid and semi-solid varieties, having a specific gravity similar to that of coal, and pitchy lustre with a black or brownish black colour. It contains from 7 to 9 per cent. of hydrogen, and sometimes a considerable proportion of oxygen and some earthy impurities. It is found in veins and beds, and has proceeded from the alteration and hardening of petroleum, owing to the loss of its more volatile ingredients.

Albertite and " *Levis Coal*," are asphaltic minerals still further altered, until they assume nearly the appearance and composition of the bituminous coals. They are found in veins or fissures, and not in beds like the true coals, and have no vegetable structure. In some altered rocks materials of this kind have been converted into anthracite and probably into graphite.

Earthy Bitumen and *Cannel Coal* are materials of this series, mixed with much earthy matter, and hardened till they resemble true coals. They are found in beds associated with the ordinary coals, and are much used in gas-making and for the distillation of coal oil.

It will be seen that the Coals and Bitumens form two parallel series, according to the amount of chemical change which they have experienced, thus :—

COAL SERIES.	BITUMEN SERIES.
Vegetable matter.	Vegetable matter.
Peat.	Petroleum.
Brown coal.	Asphaltum.
Bituminous coal.	Cannel coal.
Anthracite.	Anthracite.
Graphite.	Graphite.

25. GRAPHITE.

This substance is Carbon with its molecules arranged in a peculiar manner, constituting an allotropic form. Its crystalline form is hexagonal, in flat six-sided tables.

H.—1. to 2.   Gr.—2.

Colour black and steel grey ; streak black. Lustre metallic. Divides into thin laminæ, flexible and greasy to touch.

Graphite is probably, in most cases, a coal or asphalt, altered by heat, and in this way it is often formed accidentally in furnaces. It is largely used in making crucibles for melting metals, in coating iron castings, in lessening the friction of machinery, and in drawing and writing. Its common names of "black lead" and "plumbago" are inappropriate, as it contains no lead. The name Graphite is derived from its use in writing.

[For other minerals occurring disseminated in rocks or in veins and other repositories, the student is referred to text-books of Mineralogy.]